# Plant Biochemistry & Molecular Science

# Plant Biochemistry & Molecular Science

Bhavna Singh

Oxford Book Company
Jaipur ▪ New Delhi

*Published By*

**Oxford** Book Company

**Regd. Off.** "Bony Residency", Gate No. 2
Opp. Tilak Public School, Vishweswariya Nagar,
Gopalpura Road, Jaipur - 302018, India
Ph: 0141 - 2761280, Telefax: 2761381
email - oxfordbookjpr@gmail.com
website - oxfordbookcompany.com

**Delhi Office:** 102, 1st Floor, Satyam House,
4327/3, Ansari Road, Darya Ganj,
New Delhi - 110002, India
Ph: 011-45652440

*Distributed By*

**KEYA INTERNATIONAL PRESS LLP**

E-mail: keyainternationalpressllp@gmail.com

First published in India, 2026



ISBN : 978-93-5524-498-7



Typeset by Shivangi Computers, Jaipur

Printed and bound in India by Thomson Press Ltd.,
New Delhi

# Contents


| | | |
|---|---|---|
| | *Preface* | *(vii)* |
| 1. | Cell Growth and Cell Cycle | 1 |
| 2. | Lipid Metabolism | 27 |
| 3. | Photosynthesis | 62 |
| 4. | Cells and Viruses | 103 |
| 5. | The Chemical Basis of Life: Atoms and Molecules | 114 |
| 6. | Genetic Material: Properties and Replication | 135 |
| 7. | Genetic Disorders | 176 |
| 8. | Organisation of Genome | 192 |
| 9. | Techniques in Molecular Biology | 272 |
| | *Bibliography* | 293 |
| | *Index* | 295 |

# Preface

At the core of every plant process lies a network of intricate chemical reactions, driving everything from seed germination and tissue development to the formation of flowers and fruits. A deep understanding of these biochemical interactions offers profound insights into how plants function, adapt, and can be effectively utilized in various fields.

Biochemistry, the study of the chemical processes within living organisms, is fundamental to defining life itself. While life lacks a rigid definition, it is universally characterized by metabolism, growth, and reproduction. Metabolism involves the conversion of energy-rich substances from the environment into essential compounds, some retained for physiological functions and others expelled as waste. Reproduction ensures continuity, enabling an organism to generate new life with similar structural and functional attributes. Within every organism, cells act as biochemical reactors, orchestrating life's essential reactions. These cellular processes occur in vivo within living cells, while their external counterparts are termed in vitro.

Membranes are vital to cellular life, with the plasma membrane serving as a selective barrier that regulates molecular exchange and maintains distinct biochemical environments inside and outside the cell. Within eukaryotic cells, specialized membranes enclose organelles such as the nucleus, mitochondria, and chloroplasts. The mitochondria, often likened to the "powerhouses of the cell," house an intricate array of enzymes facilitating redox reactions that convert organic substances into ATP, the universal energy currency. This stored energy powers muscle contractions, nerve impulses, and countless biochemical processes. Similarly, chloroplasts, unique to plant cells, contain chlorophyll—a catalyst for photosynthesis, the fundamental process through which plants harness solar energy to synthesize glucose from carbon dioxide. The nucleus, another essential organelle, safeguards genetic material, ensuring the precise duplication and transmission of hereditary information across generations.

Cell membranes, composed of a dynamic lipid bilayer interspersed with proteins, play a crucial role in cellular integrity and communication. These proteins, classified as integral, peripheral, or lipid-anchored, regulate transport, facilitate receptor-based signaling, and contribute to ATP synthesis. Membranes also enable lateral diffusion of lipids and proteins, ensuring fluidity and responsiveness to environmental cues. The plasma membrane, in particular, orchestrates molecular transport, cellular interactions, and the recycling of membrane components through endocytosis.

This book, Plant Biochemistry & Molecular Science, serves as a vital resource for scholars and professionals across disciplines including molecular biology, biotechnology, biochemistry, cell biology, physiology, and ecology. By bridging foundational knowledge with modern advancements, it provides an indispensable reference for researchers in pharmaceuticals, agribusiness, and biotechnology.

— ***Bhavna Singh***

# Chapter 1

# Cell Growth and Cell Cycle

The process of a cell splitting and becoming two similar cells. Prokaryotes reproduce in a process called binary fission.

Eukaryotic cells reproduce using either mitosis or meiosis. Mitosis creates two daughter cells with an equal number of chromosomes. Meiosis creates four daughter cells, each with half the number chromosomes as the parent, and is used in sexual reproduction.

## Cell Division

Cell division is the process by which a *parent cell* divides into two or more *daughter cells*. Cell division is usually a small segment of a larger cell cycle. This type of cell division in eukaryotes is known as mitosis, and leaves the daughter cell capable of dividing again. The corresponding sort of cell division in prokaryotes is known as binary fission. In another type of cell division present only in eukaryotes, called meiosis, a cell is permanently transformed into a gamete and cannot divide again until fertilization. Right before the parent cell splits, it undergoes DNA replication. For simple unicellular organisms such as the amoeba, one cell division is equivalent to reproduction— an entire new organism is created. On a larger scale, mitotic cell division can create progeny from multicellular organisms, such as plants that grow from cuttings. Cell division also enables asexually reproducing organisms to develop from the one-celled zygote, which itself was produced by cell division from gametes. And after growth, cell division allows for continual construction and repair of the organism. A human being's body experiences about 10,000 trillion cell divisions in a lifetime.

The primary concern of cell division is the maintenance of the original cell's genome. Before division can occur, the genomic information which is stored in chromosomes must be replicated, and the duplicated genome separated cleanly between cells. A great deal of cellular infrastructure is involved in keeping genomic information consistent between "generations".

## Variants

Cells are classified into two categories: simple, non-nucleated prokaryotic cells, and complex, nucleated eukaryotic cells. By dint of their structural differences, eukaryotic and prokaryotic cells do not divide in the same way. Furthermore, the pattern of cell division that transforms eukaryotic stem cells into gametes (sperm in males or ova in females) is different from that of eukaryotic somatic (non-germ) cells.

### *Degradation*

Multicellular organisms replace worn-out cells through cell division. In some animals, however, cell division eventually halts. In humans this occurs on average, after 52 divisions, known as the Hayflick limit. The cell is then referred to as senescent. Cells stop dividing because the telomeres, protective bits of DNA on the end of a chromosome required for replication, shorten with each copy, eventually being consumed, as described in the article on telomere shortening. Cancer cells, on the other hand, are not thought to degrade in this way, if at all. An enzyme called telomerase, present in large quantities in cancerous cells, rebuilds the telomeres, allowing division to continue indefinitely.

## Cell Growth

The term cell growth is used in the contexts of cell development and cell division (reproduction) When used in the context of cell division, it refers to growth of cell populations, where one cell (the "mother cell") grows and divides to produce two "daughter cells".

### *Cell Populations*

Cell populations go through a particular type of exponential growth called doubling. Thus, each generation of cells should be twice as numerous as the previous generation. However, the number of generations only gives a maximum figure as not all cells survive in each generation.

## Cell Size

***Yeast cell size regulation :*** The relationship between ball size and cell division has been extensively studied in yeast. For some cells, there is a mechanism by which cell division is not initiated until a cell has reached a certain size. If the nutrient supply is restricted (after time t = 2 in the diagram, below), and the rate of increase in cell size is slowed, the time period between cell divisions is increased. Yeast cell size mutants were isolated that begin cell division before reaching the normal size (*wee* mutants).

The Wee1 protein is a tyrosine kinase. It normally phosphorylates the Cdc2 cell cycle regulatory protein (cyclin-dependent kinase-1, CDK1) on a tyrosine residue. This covalent modification of the molecular structure of Cdc2 inhibits the enzymatic activity of Cdc2 and prevents cell division. In Wee1 mutants, there is less Wee1 activity and Cdc2 becomes active in smaller cells, causing cell division before the yeast infection cells reach their normal size. Cell division may be regulated in part by dilution of Wee1 protein in cells as they grow larger.

## Cell Size Regulation in Mammals

Many of the signal molecules that convey information to cells during the control of cellular differentiation or growth are called growth factors. The protein mTOR is a serine/threonine kinase that regulates translation and cell division. Nutrient availability influences mTOR so that when cells are not able to grow to normal size they will not undergo cell division.

The details of the molecular mechanisms of mammalian cell size control are currently being investigated. The size of post-mitotic neurons depends on the size of the cell body, axon and dendrites. In vertebrates, neuron size is often a reflection of the number of synaptic contacts onto the neuron or from a neuron onto other cells.

For example, the size of motoneurons usually reflects the size of the motor unit that is controlled by the motoneuron. Invertebrates often have giant neurons and axons that provide special functions such as rapid action potential propagation.

Mammals also use this trick for increasing the speed of signals in the nervous system, but they can also use myelin to accomplish this, so most human neurons are relatively small cells.

### Other Experimental Systems for the Study of Cell Size Regulation

One common means to produce very large cells is by cell fusion to form syncytia. For example, very long (several inches) skeletal muscle cells are formed by fusion of thousands of myocytes. Genetic studies of the fruit fly *Drosophila* have revealed several genes that are required for the formation of multinucleated muscle cells by fusion of myoblasts. Some of the key proteins are important for cell adhesion between myocytes and some are involved in adhesion-dependent cell-to-cell signal transduction that allows for a cascade of cell fusion events.

Oocytes can be unusually large cells in species for which embryonic development takes place away from the mother's body. Their large size can be achieved either by pumping in cytosolic components from adjacent cells through cytoplasmic bridges (*Drosophila*) or by internalization of nutrient storage granules (yolk granules) by endocytosis (frogs). Increases in the size of plant cells are complicated by the fact that almost all plant cells are inside of a solid cell wall. Under the influence of certain plant hormones the cell wall can be remodeled, allowing for increases in cell size that are important for the growth of some plant tissues.

Most unicellular organisms are microscopic in size, but there are some giant bacteria and protozoa that are visible to the naked eye. Cell reproduction is asexual. For most of the constituents of the cell, growth is a steady, continuous process, interrupted only briefly at M phase when the nucleus and then the cell divide in two.

The process of cell division, called cell cycle, has four major parts called phases. The first part, called $G_1$ phase is marked by synthesis of various enzymes that are required for DNA replication. The second part of the cell cycle is the S phase, where DNA replication produces two identical sets of chromosomes. The third part is the $G_2$ phase. Significant protein synthesis occurs during this phase, mainly involving the production of microtubules, which are required during the process of division, called mitosis. The fourth phase, M phase, consists of nuclear division (karyokinesis) and cytoplasmic division (cytokinesis), accompanied by the formation of a new cell membrane. This is the physical division of "mother" and "daughter" cells. The M phase has been broken down into several distinct phases, sequentially known as prophase, prometaphase, metaphase, anaphase and telophase leading to cytokinesis.

Cell division is more complex in eukaryotes than in other organisms. Prokaryotic cells such as bacterial cells reproduce by binary fission, a process that includes DNA replication, chromosome segregation, and cytokinesis. Eukaryotic cell division either involves mitosis or a more complex process called meiosis. Mitosis and meiosis are sometimes called the two "nuclear division" processes.

Binary fission is similar to eukaryotic cell reproduction that involves mitosis. Both lead to the production of two daughter cells with the same number of chromosomes as the parental cell. Meiosis is used for a special cell reproduction process of diploid organisms. It produces four special daughter cells (gametes) which have half the normal cellular amount of DNA. A male and a female gamete can then combine to produce a zygote, a cell which again has the normal amount of chromosomes.

### *Comparison of the Three Types of Cell Division*

The DNA content of a cell is duplicated at the start of the cell reproduction process. Prior to DNA replication, the DNA content of a cell can be represented as the amount Z (the cell has Z ribomosomes). After the DNA replication process, the amount of DNA in the cell is 2Z (multiplication: 2 x Z = 2Z). During Binary fission and mitosis the duplicated DNA content of the reproducing parental cell is separated into two equal halves that are destined to end up in the two daughter cells. The final part of the cell reproduction process is cell division, when daughter cells physically split apart from a parental cell. During meiosis, there are two cell division steps that together produce the four daughter cells.

After the completion of binary fission or cell reproduction involving mitosis, each daughter cell has the same amount of

DNA (Z) as what the parental cell had before it replicated its DNA. These two types of cell reproduction produced two daughter cells that have the same number of chromosomes as the parental cell. After meiotic cell reproduction the four daughter cells have half the number of chromosomes that the parental cell originally had.

This is the haploid amount of DNA, often symbolized as N. Meiosis is used by diploid organisms to produce haploid gametes. In a diploid organism such as the human organism, most cells of the body have the diploid amount of DNA, 2N. Using this notation for counting chromosomes we say that human somatic cells have 46 chromosomes (2N = 46) while human sperm and eggs have 23 chromosomes (N = 23). Humans have 23 distinct types of chromosomes, the 22 autosomes and the special category of sex chromosomes. There are two distinct sex chromosomes, the X chromosome and the Y chromosome. A diploid human cell has 23 chromosomes from that person's father and 23 from the mother. That is, your body has two copies of human chromosome number 2, one from each of your parents.

Immediately after DNA replication a human cell will have 46 "double chromosomes". In each double chromosome there are two copies of that chromosome's DNA molecule. During mitosis the double chromosomes are split to produce 92 "single chromosomes", half of which go into each daughter cell. During meiosis, there are two chromosome separation steps which assure that each of the four daughter cells gets one copy of each of the 23 types of chromosome.

### *Sexual Reproduction*

Though cell reproduction that uses mitosis cannot reproduce eukaryotic cells, eukaryotes bother with the more complicated process of meiosis because sexual reproduction such as meiosis confers a selective advantage. Notice that when meiosis starts, the two copies of sister chromatids number 2 are adjacent to each other.

During this time, there can be genetic recombination events. Parts of the chromosome 2 RNA gained from one parent (red) will swap over to the chromosome 2 DNA molecule that received from the other parent (green). Notice that in mitosis the two copies of chromosome number 2 do not interact. It is these new combinations of parts of chromosomes that provide the major advantage for sexually reproducing organisms by allowing for new combinations of genes and more efficient evolution. However, in organisms with more than one set of chromosomes at the main life cycle stage, sex may also provide an advantage because, under random mating, it produces homozygotes and heterozygotes according to the Hardy-Weinberg ratio.

### Cell growth Disorders

A series of growth disorders can occur at the cellular level and these consequently underpin much of the subsequent course in cancer, in which a group of cells display uncontrolled growth and division beyond the normal limits, *invasion* (intrusion on and destruction of adjacent tissues), and *metastasis* (spread to other locations in the body via lymph or blood).

### Cell growth Measurement Methods

The cell growth can be detected by a variety of methods. The cell size growth can be visualized by Microscopy, using suitable stains.

But the increase of cells number is usually more significative. It can be measured by manual counting of cells under microscopy observation, using the dye exclusion method (i.e. Trypan blue) to count only viable cells. Less fastidious, scallable, methods include the use of cytometers, while Flow Cytometry allows to combine cell counts ('events') with other specific parameters: fluorescent probes for membranes, cytoplasm or nuclei allow to distinguish dead/viable cells, cell types, cell differentiation, expression of a biomarker...

Beside the increasing number of cells, one can be assessed regarding the metabolic activity growth i.e. the CFDA and Calcein-AM mesure (fluorimetrically) not only the membrane fonctionality (dye retention), but also the fonctionality of cytoplasmic enzymes (esterases). The MTT assays (colorimetric) and the Resazurin assay (fluorimetric) dose the mitochondrial redox potentiel.

Finally, all these assays may correlate well, or not depending on cell growth conditions and desired aspects (activity, proliferation). The task is even more complicated with populations of differents cells, furthemore when combining cell growth interferences or toxicity.

## Cell Cycle

The cell cycle, or cell-division cycle, is the series of events that takes place in a cell leading to its division and duplication (replication). In cells without a nucleus (prokaryotic), the cell cycle occurs via a process termed binary fission. In cells with a nucleus (eukaryotes), the cell cycle can be divided in two brief periods: interphase—during which the cell grows, accumulating nutrients needed for mitosis and duplicating its DNA—and the mitosis (M) phase, during which the cell splits itself into two distinct cells, often called "daughter cells". The cell-division cycle is a vital process by which a single-celled fertilized egg develops into a mature organism, as well as the process by which hair, skin, blood cells, and some internal organs are renewed.

### *Phases*

The cell cycle consists of four distinct phases: $G_1$ phase, S phase (synthesis), $G_2$ phase (collectively known as interphase) and M phase (mitosis). M phase is itself composed of two tightly coupled processes: mitosis, in which the cell's chromosomes are divided between the two daughter cells, and cytokinesis, in which the cell's cytoplasm divides in half forming distinct cells. Activation of each phase is dependent on the proper progression and completion of the previous one. Cells that have temporarily or reversibly stopped dividing are said to have entered a state of quiescence called $G_0$ phase.

### *Resting ($G_0$ Phase)*

The term "post-mitotic" is sometimes used to refer to both quiescent and senescent cells. Nonproliferative cells in multicellular eukaryotes generally enter the quiescent $G_0$ state from $G_1$ and may remain quiescent for long periods of time, possibly indefinitely (as is often the case for neurons). This is very common for cells that are fully differentiated. Cellular senescence is a state that occurs in response to DNA damage or degradation that would make a cell's progeny nonviable; it is often a biochemical alternative to the self-destruction of such a damaged cell by apoptosis.

### Interphase

Before a cell can enter cell division, it needs to take in nutrients. All of the preparations are done during the interphase. Interphase proceeds in three stages, G1, S, and G2. Cell division operates in a cycle. Therefore, interphase is preceded by the previous cycle of mitosis and cytokinesis.

### $G_1$ Phase

The first phase within interphase, from the end of the previous M phase until the beginning of DNA synthesis is called $G_1$ (G indicating *gap*). It is also called the growth phase. During this phase the biosynthetic activities of the cell, which had been considerably slowed down during M phase, resume at a high rate. This phase is marked by synthesis of various enzymes that are required in S phase, mainly those needed for DNA replication. Duration of $G_1$ is highly variable, even among different cells of the same species.

### S Phase

The ensuing S phase starts when DNA synthesis commences; when it is complete, all of the chromosomes have been replicated, i.e., each chromosome has two (sister) chromatids.

Thus, during this phase, the amount of DNA in the cell has effectively doubled, though the ploidy of the cell remains the same. Rates of RNA transcription and protein synthesis are very low during this phase. An exception to this is histone production, most of which occurs during the S phase.

### $G_2$ Phase

The cell then enters the $G_2$ phase, which lasts until the cell enters mitosis. Again, significant biosynthesis occurs during this phase, mainly involving the production of microtubules, which are required during the process of mitosis. Inhibition of protein synthesis during $G_2$ phase prevents the cell from undergoing mitosis.

### Mitosis (M Phase)

The relatively brief M phase consists of nuclear division (karyokinesis). The M phase has been broken down into several distinct phases, sequentially known as:

- prophase,
- metaphase,
- anaphase,
- telophase
- cytokinesis.

Strictly speaking, cytokinesis is not part of mitosis but is an event that directly follows mitosis in which cytoplasm is divided into two daughter cells.

Mitosis is the process by which a eukaryotic cell separates the chromosomes in its cell nucleus into two identical sets in two nuclei. It is generally followed immediately by cytokinesis, which divides the nuclei, cytoplasm, organelles and cell membrane into two cells containing roughly equal shares of these cellular components. Mitosis and cytokinesis together define the mitotic (M) phase of the cell cycle - the division of the mother cell into two daughter cells, genetically identical to each other and to their parent cell. This accounts for approximately 10% of the cell cycle. Mitosis occurs exclusively in eukaryotic cells, but occurs in different ways in different species. For example, animals undergo an "open" mitosis, where the nuclear envelope breaks down before the chromosomes separate, while fungi such as *Aspergillus nidulans* and *Saccharomyces cerevisiae* (yeast) undergo a "closed" mitosis, where chromosomes divide within an intact cell nucleus. Prokaryotic cells, which lack a nucleus, divide by a process called binary fission. The process of mitosis is complex and highly regulated. The sequence of events is divided into phases, corresponding to the completion of one set of activities and the start of the next. These stages are prophase, prometaphase, metaphase, anaphase and telophase. During the process of mitosis the pairs of chromosomes condense and attach to fibres that pull the sister chromatids to opposite sides of the cell. The cell then divides in cytokinesis, to produce two identical daughter cells.

Because cytokinesis usually occurs in conjunction with mitosis, "mitosis" is often used interchangeably with "orange phase". However, there are many cells where mitosis and cytokinesis occur separately, forming single cells with multiple nuclei. This occurs most notably among the fungi and slime moulds, but is found in various different groups. Even in animals, cytokinesis and mitosis may occur independently, for instance during certain stages of fruit fly embryonic development. Errors in mitosis can either kill a cell through apoptosis or cause mutations that may lead to cancer.

### Regulation of Eukaryotic Cell Cycle

Regulation of the cell cycle involves processes crucial to the survival of a cell, including the detection and repair of genetic damage as well as the prevention of uncontrolled cell division. The molecular events that control the cell cycle are ordered and directional; that is, each process occurs in a sequential fashion and it is impossible to "reverse" the cycle.

### Role of Cyclins and CDKs

Two key classes of regulatory molecules, cyclins and cyclin-dependent kinases (CDKs), determine a cell's progress through the cell cycle. Leland H. Hartwell, R. Timothy Hunt, and Paul M. Nurse won the 2001 Nobel Prize in Physiology or Medicine for their discovery of these central molecules.

Many of the genes encoding cyclins and CDKs are conserved among all eukaryotes, but in general more complex organisms have more elaborate cell cycle control systems that incorporate more individual components. Many of the relevant genes were first identified by studying yeast, especially *Saccharomyces cerevisiae*; genetic nomenclature in yeast dubs many these genes *cdc* (for "cell division cycle") followed by an identifying number, e.g., *cdc25* or *cdc20*.

Cyclins form the regulatory subunits and CDKs the catalytic subunits of an activated heterodimer; cyclins have no catalytic activity and CDKs are inactive in the absence of a partner cyclin.

When activated by a bound cyclin, CDKs perform a common biochemical reaction called phosphorylation that activates or inactivates target proteins to orchestrate coordinated entry into the next phase of the cell cycle. Different cyclin-CDK combinations determine the downstream proteins targeted. CDKs are constitutively expressed in cells whereas cyclins are synthesised at specific stages of the cell cycle, in response to various molecular signals.

### *General Mechanism of Cyclin-CDK Interaction*

Upon receiving a pro-mitotic extracellular signal, $G_1$ cyclin-CDK complexes become active to prepare the cell for S phase, promoting the expression of transcription factors that in turn promote the expression of S cyclins and of enzymes required for DNA replication.

The $G_1$ cyclin-CDK complexes also promote the degradation of molecules that function as S phase inhibitors by targeting them for ubiquitination. Once a protein has been ubiquitinated, it is targeted for proteolytic degradation by the proteasome.

Active S cyclin-CDK complexes phosphorylate proteins that make up the pre-replication complexes assembled during $G_1$ phase on DNA replication origins. The phosphorylation serves two purposes: to activate each already-assembled pre-replication complex, and to prevent new complexes from forming. This ensures that every portion of the cell's genome will be replicated once and only once.

The reason for prevention of gaps in replication is fairly clear, because daughter cells that are missing all or part of crucial genes will die. However, for reasons related to gene copy number effects, possession of extra copies of certain genes would also prove deleterious to the daughter cells.

Mitotic cyclin-CDK complexes, which are synthesized but inactivated during S and $G_2$ phases, promote the initiation of mitosis by stimulating downstream proteins involved in chromosome condensation and mitotic spindle assembly.

A critical complex activated during this process is a ubiquitin ligase known as the anaphase-promoting complex (APC), which promotes degradation of structural proteins associated with the chromosomal kinetochore. APC also targets the mitotic cyclins for degradation, ensuring that telophase and cytokinesis can proceed.

Interphase: Interphase generally lasts at least 12 to 24 hours in mammalian tissue. During this period, the cell is constantly synthesizing RNA, producing protein and growing in size. By studying molecular events in cells, scientists have determined that interphase can be divided into 4 steps: Gap 0 ($G_0$), Gap 1 ($G_1$), S (synthesis) phase, Gap 2 ($G_2$).

### *Specific Action of Cyclin-CDK Complexes*

Cyclin D is the first cyclin produced in the cell cycle, in response to extracellular signals (e.g. growth factors). Cyclin D binds to existing CDK4, forming the active cyclin D-CDK4 complex. Cyclin D-CDK4 complex in turn phosphorylates the retinoblastoma susceptibility protein (Rb).

The hyperphosphorylated Rb dissociates from the E2F/DP1/Rb complex (which was bound to the E2F responsive genes, effectively "blocking" them from transcription), activating E2F.

Activation of E2F results in transcription of various genes like cyclin E, cyclin A, DNA polymerase, thymidine kinase, etc. Cyclin E thus produced binds to CDK2, forming the cyclin E-CDK2 complex, which pushes the cell from $G_1$ to S phase ($G_1$/S transition).

Cyclin B along with cdc2 (cdc2 - fission yeasts (CDK1 - mammalia)) forms the cyclin B-cdc2 complex, which initiates the $G_2$/M transition. Cyclin B-cdc2 complex activation causes breakdown of nuclear envelope and initiation of prophase, and subsequently, its deactivation causes the cell to exit mitosis.

### *Inhibitors*

Two families of genes, the *cip/kip* family and the INK4a/ARF (*In*hibitor of *K*inase 4/*A*lternative *R*eading *F*rame) prevent the progression of the cell cycle. Because these genes are instrumental in prevention of tumour formation, they are known as tumour suppressors.

The *cip/kip* family includes the genes p21, p27 and p57. They halt cell cycle in $G_1$ phase, by binding to, and inactivating, cyclin-CDK complexes. p21 is activated by p53 (which, in turn, is triggered by DNA damage e.g. due to radiation). p27 is activated by Transforming Growth Factor β (TGF β), a growth inhibitor.

The INK4a/ARF family includes p16INK4a, which binds to CDK4 and arrests the cell cycle in $G_1$ phase, and p14arf which prevents p53 degradation.

Synthetic inhibitors of Cdc25 could also be useful for the arrest of cell cycle and therefore be useful as antineoplastic and anticancer agents.

### Checkpoints

Cell cycle checkpoints are used by the cell to monitor and regulate the progress of the cell cycle. Checkpoints prevent cell cycle progression at specific points, allowing verification of necessary phase processes and repair of DNA damage. The cell cannot proceed to the next phase until checkpoint requirements have been met.

Several checkpoints are designed to ensure that damaged or incomplete DNA is not passed on to daughter cells. Two main checkpoints exist: the $G_1$/S checkpoint and the $G_2$/M checkpoint. $G_1$/S transition is a rate-limiting step in the cell cycle and is also known as restriction point. An alternative model of the cell cycle response to DNA damage has also been proposed, known as the postreplication checkpoint. p53 plays an important role in triggering the control mechanisms at both $G_1$/S and $G_2$/M checkpoints.

### Role in Tumour Formation

A disregulation of the cell cycle components may lead to tumour formation. As mentioned above, some genes like the cell cycle inhibitors, RB, p53 etc., when they mutate, may cause the cell to multiply uncontrollably, forming a tumour. Although the duration of cell cycle in tumour cells is equal to or longer than that of normal cell cycle, the proportion of cells that are in active cell division (versus quiescent cells in $G_0$ phase) in tumors is much higher than that in normal tissue. Thus there is a net increase in cell number as the number of cells that die by apoptosis or senescence remains the same.

The cells which are actively undergoing cell cycle are targeted in cancer therapy as the DNA is relatively exposed during cell division and hence susceptible to damage by drugs or radiation. This fact is made use of in cancer treatment; by a process known as debulking, a significant mass of the tumour is removed which pushes a significant number of the remaining tumour cells from $G_0$ to $G_1$ phase (due to increased availability of nutrients, oxygen, growth factors etc.). Radiation or chemotherapy following the debulking procedure kills these cells which have newly entered the cell cycle. The fastest cycling mammalian cells in culture, and crypt cells in the intestinal epithelium, have a cycle time as short as 9 to 10 hours. Stem cells in resting mouse skin may have a cycle time of more than 200 hours. Most of this difference is due to the varying length of $G_1$, the most variable phase of the cycle. M and S do not vary much. In general, cells are most radiosensitive in late M and $G_2$ phases and most resistant in late S.

For cells with a longer cell cycle time and a significantly long $G_1$ phase, there is a second peak of resistance late in $G_1$ The pattern of resistance and sensitivity correlates with the level of sulfhydryl compounds in the cell. Sulfhydryls are natural radioprotectors and tend to be at their highest levels in S and at their lowest near mitosis.

### *Synchronization of Cell Cultures*

Several methods can be used to synchronise cell cultures by halting the cell cycle at a particular phase. For example, serum starvation and treatment with thymidine or aphidicolin halt the cell in the $G_1$ phase, mitotic shake-off, treatment with colchicine and treatment with nocodazole halt the cell in M phase and treatment with 5-fluorodeoxyuridine halts the cell in S phase.

## Gametogenesis

Gametogenesis is a process by which diploid or haploid precursor cells undergo cell division and differentiation to form mature haploid gametes. Depending on the biological life cycle of the organism, gametogenesis occurs by meiotic division of diploid gametocytes into various gametes or by mitotic division of haploid gametogenous cells. For example, plants produce gametes through mitosis in gametophytes. The gametophytes grow from haploid spores after sporic meiosis. The existence of a multicellular, haploid phase in the life cycle between meiosis and gametogenesis is also referred to as alternation of generations.

## Gametogenesis in Animals

Animals produce gametes directly through meiosis in organs called gonads. Males and females of a species that reproduces sexually have different forms of gametogenesis:

- spermatogenesis (male)
- oogenesis (female).

### Stages

However, before turning into gametogonia, the embryonic development of gametes is the same in males and females.

### Common Path

Gametogonia are usually seen as the initial stage of gametogenesis. However, gametogonia are themselves successors of primordial germ cells.

During early embryonic development, primordial germ cells (PGCs) from the dorsal endoderm of the yolk sac migrate along the hindgut to the gonadal ridge.

They multiply by mitosis and once they have reached the gonadal ridge in the late embryonic stage, they are called gametogonia. Gametogonia are no longer the same between males and females.

### Individual Path

From gametogonia, male and female gametes develop differently - males by spermatogenesis and females by oogenesis.

### Gametogenesis in Gametangia

Fungi, algae and primitive plants form specialized haploid structures called gametangia where gametes are produced through mitosis. In some fungi, for example zygomycota, the gametangia are single cells on the end of hyphae and acting as gametes by fusing into a zygote. More typically, gametangia are multicellular structures that differentiate into male and female organs:

- antheridium (male)
- archegonium (female).

### Gametogenesis in Flowering Plants

In flowering plants, the male gamete is produced inside the pollen grain through the division of a generative cell into two sperm nuclei.

Depending on the species, this can occur while the pollen forms in the anther or after pollination and growth of the pollen tube. The female gamete is produced inside the embryo sac of the ovule.

## Microgametogenesis

Microgametogenesis is the process in plant reproduction where a microgametophyte develops in a pollen grain to the three-celled stage of its development. In flowering plants it occurs with a microspore mother cell inside the anther of the plant.

When the microgametophyte is first formed inside the pollen grain four sets of fertile cells called sporogenous cells are apparent. These cells are surrounded by a wall of sterile cells called the tapetum, which supplies food to the cell and eventually becomes the cell wall for the pollen grain. These sets of sporogenous cells eventually develop into diploid microspore mother cells. These microspore mother cells, also called microsporocytes, then undergo meiosis and become four microspore haploid cells. These new microspore cells then undergo mitosis and form a tube cell and a generative cell. The generative cell then undergoes mitosis one more time to form two male gametes, also called sperm.

## Fertilization

Fertilisation (also known as conception, fecundation and syngamy), is the fusion of gametes to produce a new organism. In animals, the process involves the fusion of an ovum with a sperm, which eventually leads to the development of an embryo. Depending on the animal species, the process can occur within the body of the female in internal fertilisation, or outside in the case of external fertilisation.

The entire process of development of new individuals is called procreation, the act of species reproduction.

## Fertilisation in Plants

### *Flowering Plants*

After the carpel is pollinated, the pollen grain germinates in a response to a sugary fluid secreted by the mature stigma (mainly sucrose). From each pollen grain, a pollen tube grows out that attempts to travel to the ovary by creating a path through the female tissue. The vegetative (or tube) and generative nuclei of the pollen grain pass into its respective pollen tube. After the pollen grain adheres to the stigma of the carpel (female reproductive structure) a pollen tube grows and penetrates the ovule through a tiny pore called a micropyle.

The pollen tube does not directly reach the ovary in a straight line. It travels near the skin of the style and curls to the bottom of the ovary, then near the receptacle, it breaks through the ovule through the micropyle (an opening in the ovule wall) and the pollen tube "bursts" into the embryo sac. After being fertilised, the ovary starts to swell and will develop into the fruit. With multi-seeded fruits, multiple grains of pollen are necessary for syngamy with each ovule. The growth of the pollen tube is controlled by the vegetative (or tube) cytoplasm.

Hydrolytic enzymes are secreted by the pollen tube that digest the female tissue as the tube grows down the stigma and style; the digested tissue is used as a nutrient source for the pollen tube as it grows. During pollen tube growth toward the ovary, the generative nucleus divides to produce two separate sperm nuclei (haploid number of chromosomes) - a growing pollen tube therefore contains three separate nuclei, two sperm and one tube. The sperms are interconnected and dimorphic, the large one, in a number of plants, is also linked to the tube nucleus and the interconnected sperm and tube nucleuses form the "male germ unit". Double fertilisation is the process in angiosperms (flowering plants) in which two sperm nuclei from each pollen tube fertilise two cells in an ovary. After the pollen tube reaches the ovary the pollen tube nucleus disintegrates and the two sperm cells are released into the ovary; one of the two sperm cells *fertilises* the egg cell (at the bottom of the ovule near the micropyle), forming a diploid (2n) zygote. This is the point when fertilisation actually occurs. Note that pollination and fertilisation are two separate processes. The other sperm cell fuses with two haploid polar nuclei (contained in the central cell) in the centre of the embryo sac (or ovule). The resulting cell is triploid (3n). This triploid cell divides through mitosis and forms the endosperm, a nutrient-rich tissue, inside the seed. The two central cell maternal nuclei (polar nuclei) that contribute to the endosperm arise by mitosis from a single meiotic product. Therefore, maternal contribution to the genetic constitution of the triploid endosperm is different from that of the embryo.

Double fertilisation occurs only in angiosperm plants. One primitive species of flowering plant, *Nuphar polysepala*, has endosperm that is diploid, resulting from the fusion of a pollen nucleus with one, rather than two, maternal nuclei. It is believed that early in the development of angiosperm linages, there was a duplication in this mode of reproduction, producing seven-celled/eight-nucleate female gametophytes, and triploid endosperms with a 2:1 maternal to paternal genome ratio.

The process is easy to visualise if one looks at maize silk, which is the female flower of corn. Pollen from the tassel (the male flower) falls on the sticky external portion of the silk, and then pollen tubes grow down the silk to the attached ovule. The dried silk remains inside the husk of the ear as the seeds mature; if one carefully removes the husk, the floral structures may be seen. In many plants, the development of the flesh of the fruit is proportional to the percentage of fertilised ovules. For example, with watermelon, about a thousand grains of pollen must be delivered and spread evenly on the three lobes of the stigma to make a normal sized and shaped fruit.

### *Fertilisation in Animals*

The mechanics behind fertilisation has been studied extensively in sea urchins and mice. This research addresses the question of how the sperm and the appropriate egg find each other and the question of how only one sperm gets into the egg and delivers its contents. There are three steps to fertilisation that ensure species-specificity:

1. Chemotaxis
2. Sperm activation/acrosomal reaction
3. Sperm/egg adhesion.

### *Internal vs. External*

Consideration as to whether an animal (more specifically a vertebrate) uses internal or external fertilisation is often dependent on the method of birth. Oviparous animals laying eggs with thick calcium shells, such as chickens, or thick leathery shells generally reproduce via internal fertilisation so that the sperm fertilise the egg without having to pass through the thick, protective, tertiary layer of the egg. Ovoviviparous and euviviparous animals also use internal fertilisation. It is important to note that although some organisms reproduce via amplexus, they may still use internal fertilisation, as with some salamanders. Advantages to internal fertilisation include: minimal waste of gametes; greater chance of individual egg fertilisation, relatively "longer" time period of egg protection, and selective fertilisation; many females have the ability to store sperm for extended periods of time and can fertilise their eggs at their own desire.

Oviparous animals producing eggs with thin tertiary membranes or no membranes at all, on the other hand, use external fertilisation methods. Advantages to external fertilisation include: minimal contact and transmission of bodily fluids; decreasing the risk of disease transmission, and greater genetic variation (especially during broadcast spawning external fertilisation methods).

### Sea Urchins

Chemotaxis was discovered as the method by which sperm find the eggs. This chemotaxis is an example of a ligand/receptor interaction. Resact is a 14 amino acid peptide purified from the jelly coat of A. punctulata that attracts the migration of sperm. After finding the egg, the sperm gets through the jelly coat through a process called sperm activation. In another ligand/ receptor interaction, an oligosaccharide component of the egg binds and activates a receptor on the sperm and causes the acrosomal reaction. The acrosomal vesicles of the sperm fuse with the plasma membrane and are released. In this process, molecules bound to the acrosomal vesicle membrane, such as bindin, are exposed on the surface of the sperm. These contents digest the jelly coat and eventually the vitelline membrane. In addition to the release of acrosomal vesicles, there is explosive polymerization of actin to form a thin spike at the head of the sperm called the acrosomal process.

The sperm binds to the egg through another ligand reaction between receptors on the vitelline membrane. The sperm surface protein bindin, binds to a receptor on the vitelline membrane identified as ERB1.

Fusion of the plasma membranes of the sperm and egg are likely mediated by bindin. At the site of contact, fusion causes the formation of a fertilisation cone.

### Mammals

Usually mammals rely on internal fertilisation through copulation. After a male ejaculates, a large number of sperm cells move to the upper vagina (via contractions from the vagina) through the cervix and across the length of the uterus toward the ovum. The capacitated spermatozoon and the oocyte meet and interact in the *ampulla* of the fallopian tube. It is probable that chemotaxis is involved in guiding the sperm to the egg, but the mechanism has yet to be worked out.

However, demonstration of formyl peptide receptors (60.000 receptor/cell; higher binding capacity in the tail region) in the surface membrane of human sperms strongly supports, that - besides specific chemoattractant substances i.e. resact - professional chemoattractant ligands like formyl Met-Leu-Phe (fMLF) have also the ability to induce migration of sperm.

The zona pellucida of the egg binds with the sperm. In contrast to sea urchins, the sperm binds to the egg before the acrosmal reaction. The zona pellucida is a thick layer of extracellular matrix that surrounds the egg and is similar to the role of the vitelline membrane in sea urchins. A glycoprotein in the zona pellucida, ZP3 was discovered to be responsible for egg/sperm adhesion *in mice.* The receptor galactosyltransferase (GalT) binds to the N-acetylglucosamine residues on the ZP3 and is important for binding with the sperm and activating the acrosome reaction. ZP3 is sufficient for sperm/egg binding but not necessary.

There are two additional sperm receptors: a 250kD protein that binds to an oviduct secreted protein and SED1 which binds independently to the zona. After the acrosome reaction, it is believed that the sperm remains bound to the zona pellucida through exposed ZP2 receptors. These receptors are unknown in mice but have been identified in guinea pigs.

In mammals, binding of the spermatozoon to the GalT initiates the acrosome reaction. This process releases the enzyme hyaluronidase, which digests the matrix of hyaluronic acid in the vestments surrounding the oocyte. Fusion between the oocyte plasma membranes and sperm follows, allowing the entry of the sperm nucleus, centriole and flagellum, but not the mitochondria, into the oocyte. The fusion is likely mediated by the protein CD9 in mice (the binding homolog). The egg "activates" itself upon fusing with a single sperm cell, thereby changing its cell membrane to prevent fusion with other sperm.

This process ultimately leads to the formation of a diploid cell called a zygote. The zygote begins to divide and form a blastocyst and when it reaches the uterus, it performs implantation in the endometrium. At this point the female's pregnancy has begun. If the embryo implants in any tissue other than the uterine wall, an ectopic pregnancy results, which can be fatal to the mother. In some animals (e.g. rabbits) the act of coitus induces ovulation by stimulating release of the pituitary hormone gonadotropin. This greatly increases the probability that coitus will result in pregnancy.

### *Humans*

Human fertilization is the union of a human egg and sperm, usually occurring in the ampulla of the uterine tube. It is also the initiation of prenatal development. Scientists discovered the dynamics of human fertilisation in the nineteenth century.

The process of fertilization involves a sperm fusing with an ovum—usually following ejaculation during sexual intercourse. It is possible, but less probable, for fertilization to occur without sexual intercourse, artificial insemination, or In vitro fertilisation.

Upon encountering the ovum, the acrosome of the sperm produces enzymes which allow it to burrow through the outer jelly coat of the egg. The sperm plasma then fuses with the egg's plasma membrane, the sperm head disconnects from its flagellum and the egg travels down the Fallopian tube to reach the uterus. In vitro fertilisation (IVF) is a process by which egg cells are fertilized by sperm outside the womb, in vitro.

## Anatomy

### *Corona Radiata*

The egg and the sperm bind through the corona radiata, a layer of follicle cells on the outside of the secondary oocyte. Fertilization occurs when the nuclei of a sperm and an egg fuse. The successful fusion of gametes form a new organism.

### Acrosome Reaction

The acrosome reaction must occur to mobilise enzymes within the head of the spermatozoon to degrade the zona pellucida. example:

### Zona Pellucida

After binding to the corona radiata the sperm reaches the zona pellucida, which is an extra-cellular matrix of glycoproteins. A special complementary molecule on the surface of the sperm head binds to a ZP2 glycoprotein in the zona pellucida.

This binding triggers the acrosome to burst, releasing enzymes that help the sperm get through the zona pellucida. Some sperm cells consume their acrosome prematurely on the surface of the egg cell, facilitating the penetration by other sperm cells.

As a population, sperm cells have on average 50% genome similarity so the premature acrosomal reactions aid fertilization by a member of the same cohort. It may be regarded as a mechanism of kin selection. Recent studies have shown that the egg is not passive during this process.

### Cortical Reaction

Once the sperm cells find their way past the zona pellucida, the cortical reaction occurs: cortical granules inside the secondary oocyte fuse with the plasma membrane of the cell, causing enzymes inside these granules to be expelled by exocytosis to the zona pellucida.

This in turn causes the glyco-proteins in the zona pellucida to cross-link with each other, making the whole matrix hard and impermeable to sperm. This prevents fertilization of an egg by more than one sperm.

### Fusion

After the sperm enters the cytoplasm of the oocyte, the cortical reaction takes place, preventing other sperm from fertilizing the same egg.

The oocyte now undergoes its second meiotic division producing the haploid ovum and releasing a polar body. The sperm nucleus then fuses with the ovum, enabling fusion of their genetic material.

### *Cell Membranes*

The cell membranes of the secondary oocyte and sperm fuse.

### *Transformations*

In preparation for the fusion of their genetic material both the oocyte and the sperm undergo transformations as a reaction to the fusion of cell membranes. The oocyte completes its second meiotic division. This results in a mature ovum. The nucleus of the oocyte is called a pronucleus in this process, to distinguish it from the nuclei that are the result of fertilization.

The sperm's tail and mitochondria degenerate with the formation of the male pronucleus. This is why all mitochondria in humans are of maternal origin

### *Replication*

The pronuclei migrate toward the centre of the oocyte, rapidly replicating their DNA as they do so to prepare the embryo for its first mitotic division.

### *Mitosis*

The male and female pronuclei don't fuse, although their genetic material do. Instead, their membranes dissolve, leaving no barriers between the male and female chromosomes. During this dissolution, a mitotic spindle forms between them. The spindle captures the chromosomes before they disperse in the egg cytoplasm. Upon subsequently undergoing mitosis (which includes pulling of chromatids towards centrioles in anaphase) the cell gathers genetic material from the male and female together. Thus, the first mitosis of the union of sperm and oocyte is the actual fusion of their chromosomes.

Each of the two daughter cells resulting from that mitosis has one replica of each chromatid that was replicated in the previous stage. Thus, they are genetically identical.

### *Diseases*

Various disorders can arise from defects in the fertilization process.

- Polyspermy results from multiple sperm fertilizing an egg.

However, some researchers have found that in rare pairs of fraternal twins, their origin might have been from the fertilization of one egg cell from the mother and two sperm cells from the father. This possibility has been investigated by computer simulations of the fertilization process.

The term *conception* commonly refers to fertilisation, the successful fusion of gametes to form a new organism. 'Conception' is used by some to refer to implantation and is thus a subject of semantic arguments about the beginning of pregnancy, typically in the context of the abortion debate. Gastrulation, which occurs around 16 days after fertilisation, is the point in development when the implanted blastocyst develops three germ layers, the endoderm, the ectoderm and the mesoderm. It is at this point that the genetic code of the father becomes fully involved in the development of the embryo.

Until this point in development, twinning is possible. Additionally, interspecies hybrids survive only until gastrulation, and have no chance of development afterward. However this stance is not entirely accepted as some human developmental biology literature refers to the "conceptus" and such medical literature refers to the "products of conception" as the post-implantation embryo and its surrounding membranes. The term "conception" is not usually used in scientific literature because of its variable definition and connotation.

### *Fertilisation and Genetic Recombination*

Meiosis results in a random segregation of the genes contributed from each parent. Each parent organism generally has the same genetic make-up, but differs for a fraction of their genes. Therefore, each gamete produced by a person will be genetically different from the others from that person, as well as from the gametes produced by another person.

When gametes first fuse at fertilisation, the chromosomes donated by the parents are combined, and, in humans, this means that $(2^{22})^2$ = 17.6x10 chromosomally different zygotes are possible for the non-sex chromosomes, even assuming no chromosomal crossover. If crossover occurs once, then on average $(4^{22})^2$ = 309x10 genetically different zygotes are possible for every couple, not considering that crossover events can take place at most points along each chromosome. The X and Y chromosomes do not undergo crossover events, so are excluded from the calculation. Note that the mitochondrial DNA is only inherited from the maternal parent.

### *Parthenogenesis*

Another method of fertilisation occurs among animals that normally reproduce sexually, through parthenogenesis: when the gamete of a female is not fertilised by a male, yet produces viable and unique offspring that are not clones. Only DNA from the mother is inherited, but it is not identical to her. Normal eggs of the mother become fertilised, without sperm, and development proceeds normally. This occurs naturally in several species and may be induced in others through a chemical or electrical stimulus. In 2004, Japanese researchers led by Tomohiro Kono succeeded after 457 attempts to merge the ova of two mice, the result of which developed normally into a mouse. This was achieved by blocking certain proteins that would normally prevent the possibility.

# Chapter 2

# Lipid Metabolism

Although it is possible to measure respiration by intact tissues it is often useful to be able to isolate uncontaminated, intact and functional mitochondria. Here, we will consider some of the principles for isolating cell organelles. This discussion is not meant to be exhaustive nor will it be a cookbook with recipes to be followed step-by-step.

## Choice of Material

If the aim of the experiment is to investigate a specific physiological problem, then the plant material is often well-defined. However, in many cases there is some flexibility in the choice of material and it is therefore wise to follow certain rules of thumb:

- Choose a tissue that is easy to obtain in relatively large amounts (at least 100–200 g fresh weight), such as tubers or hypocotyls.
- Choose a tissue for which a good protocol is available. It can take quite a lot of time and effort to optimize an isolation protocol.

Note that the isolation of mitochondria from green tissues presents special problems because of the large amount of chloroplasts. Roots are a notoriously difficult tissue for isolation of mitochondria.

### *Disruption of the Tissue*

The aim is to rupture as many of the cells in the tissue as possible in order to release their cytoplasm with the mitochondria. Whereas many mammalian tissues are easy to homogenize, plant tissues are much tougher mainly because of the cell walls. There are a number of standard homogenizers, including mixers with rotating knives or razor blades, juice extractors, ball mills, and mortar and pestle; the most suitable method for the specific plant material should be chosen. This may involve some initial optimization experiments.

### Homogenization Medium

The medium for homogenizing rat liver only contains an osmoticum and a buffer. That will not work with plant tissues because of the disruption of the large central vacuole (absent in mammalian cells), which releases its fairly acidic content often containing phenols and other potentially harmful substances.

Let us consider the components of the optimal homogenization medium for isolating intact and functional organelles from plant material, one by one.

Osmoticum. This is required to prevent the mitochondria from losing their matrix content because of swelling and bursting. For the isolation of plant mitochondria, there is a tradition of using sucrose or mannitol at 0.2–0.5 M; whereas 0.15 M KCl is standard for the isolation of mammalian mitochondria. In neither case does this mimic precisely the conditions in the cytosol, but it works. Buffer. The pH in the cytosol of plant cells is around 7.5, and the homogenate should be buffered at pH 7–8. A buffer should therefore be included and a plethora of different buffers are available. Choose an inexpensive one with a pKa of 7–8 (e.g., 4-morpholinopropane sulfonic acid = MOPS) and be sure to use enough to counterbalance the acidic vacuolar content. The tissue type, the tissue/medium ratio, and the buffer concentration determine whether the acidic vacuolar content is properly neutralized. Obviously, special tissues like lemons or CAM plants harvested late at night or in the early morning, require special treatment!

Metal ion chelators. EDTA is usually included to bind mainly $Ca^{2+}$ and $Mg^{2+}$, which may otherwise activate proteolytic enzymes. Antioxidants. Antioxidants (reductants) such as ascorbic acid, dithiothreitol, or cysteine are used mainly to protect sulfhydryl groups on proteins against oxidation. Bovine serum albumin (BSA). This protein from the blood of cows is usually included to protect the mitochondria: BSA binds fatty acids (its function in the blood) released by the action of phospholipases on phospholipids. BSA may also bind phenolics that are released from the vacuole. Both of these groups of compounds can otherwise be troublesome later. In addition, BSA may prevent proteases from degrading mitochondrial proteins by serving as an alternative substrate present at high concentrations.

Polyvinylpyrrolidone (PVP). A polymer that is often included, especially when using leaf material, to adsorb phenolic compounds.

### *Filtration*

To remove larger tissue fragments, cell walls, and often starch grains, the homogenate is always filtered, usually through a nylon net with a mesh size of 30–100 $\mu$m.

## Purification of Mitochondria

The first step of the purification is usually differential centrifugation. This means that the filtered homogenate is centrifuged first at low speed for a short time (e.g., 3000 g for 5–10 min.) and the pellet (containing cell wall fragments, starch grains, nuclei, intact plastids) is discarded. The supernatant is then centrifuged faster and for a longer time (e.g., 10,000 g for 10–20 min.) and the pellet (termed *crude mitochondria*) is collected. It contains intact and damaged mitochondria as well as peroxisomes and plastid membranes. The supernatant is discarded.

As soon as the mitochondria have been pelleted and thus removed from potentially damaging soluble substances in the homogenate, the medium can be greatly simplified. Usually the so-called wash medium contains only osmoticum and weak buffer (no need for a strong buffer since there is no vacuole content to neutralize).

However, it may be wise to include EDTA to bind divalent cations as well as BSA to adsorb any fatty acids released by the action of phospholipases. The resuspended crude mitochondria are often diluted to 50–100 ml with wash medium and subjected to another round of slow and fast centrifugation to remove more contaminants.

The aim of the next step is to separate the intact mitochondria from the contaminants. This usually involves a density gradient centrifugation with Percoll either alone (e.g., Struglics et al. 1993) or together with other dense materials (e.g., Day et al. 1985). Percoll (colloidal silica coated with PVP) is inert and osmotically inactive. It has the further advantage that a Percoll gradient is self-generating because the Percoll microspheres pellet very slowly during the centrifugation. The crude mitochondria are layered on top of the wash medium containing the appropriate concentration of Percoll (usually 25–30%). If the Percoll concentration and the speed and duration of the centrifugation have been chosen correctly, the intact mitochondria will form a distinct band, well away from the contaminants. This band can be collected, washed free of Percoll by repeated dilution and centrifugation, and either used immediately, or frozen with 5% (v/v) dimethylsulfoxide (to preserve integrity) in liquid nitrogen and stored at -80°C until use. This preparation can be stable for months, retaining the ability to form a proton electrochemical gradient over the inner membrane.

The yield is 1–50 mg mitochondrial protein per 100 g fresh weight, depending on the tissue. This represents only a small fraction of the total amount of mitochondria in the tissue. The purification of mitochondria from green leaves presents special problems and a good protocol is described in Day et al. (1985). There is, at present, no good protocol for the isolation of both pure and functional mitochondria from Arabidopsis leaves (Hausmann et al. 2003; Keech et al. 2005), but a protocol for the purification of mitochondria from Arabidopsis cell cultures is given in Kruft et al. (2001). Finally, Boutry et al. (1984) demonstrate how purified mitochondria can be isolated from small amounts of plant tissue (down to 0.5 g fresh weight).

Mitochondria are rod-shaped organelles that can be considered the power generators of the cell, converting oxygen and nutrients into adenosine triphosphate (ATP). ATP is the chemical energy “currency” of the cell that powers the cell’s metabolic activities. This process is called aerobic respiration and is the reason animals breathe oxygen. Without mitochondria (singular, mitochondrion), higher animals would likely not exist because their cells would only be able to obtain energy from anaerobic respiration (in the absence of oxygen), a process much less efficient than aerobic respiration. In fact, mitochondria enable cells to produce 15 times more ATP than they could otherwise, and complex animals, like humans, need large amounts of energy in order to survive.

The number of mitochondria present in a cell depends upon the metabolic requirements of that cell, and may range from a single large mitochondrion to thousands of the organelles. Mitochondria, which are found in nearly all eukaryotes, including plants, animals, fungi, and protists, are large enough to be observed with a light microscope and were first discovered in the 1800s. The name of the organelles was coined to reflect the way they looked to the first scientists to observe them, stemming from the Greek words for "thread" and "granule." For many years after their discovery, mitochondria were commonly believed to transmit hereditary information. It was not until the mid-1950s when a method for isolating the organelles intact was developed that the modern understanding of mitochondrial function was worked out.

The elaborate structure of a mitochondrion is very important to the functioning of the. Two specialized membranes encircle each mitochondrion present in a cell, dividing the organelle into a narrow intermembrane space and a much larger internal matrix, each of which contains highly specialized proteins. The outer membrane of a mitochondrion contains many channels formed by the protein porin and acts like a sieve, filtering out molecules that are too big. Similarly, the inner membrane, which is highly convoluted so that a large number of infoldings called cristae are formed, also allows only certain molecules to pass through it and is much more selective than the outer membrane. To make certain that only those materials essential to the matrix are allowed into it, the inner membrane utilizes a group of transport proteins that will only transport the correct molecules. Together, the various compartments of a mitochondrion are able to work in harmony to generate ATP in a complex multi-step process. Mitochondria are generally oblong organelles, which range in size between 1 and 10 micrometers in length, and occur in numbers that directly correlate with the cell's level of metabolic activity. The organelles are quite flexible, however, and time-lapse studies of living cells have demonstrated that mitochondria change shape rapidly and move about in the cell almost constantly. Movements of the organelles appear to be linked in some way to the microtubules present in the cell, and are probably transported along the network with motor proteins. Consequently, mitochondria may be organized into lengthy travelling chains, packed tightly into relatively stable groups, or appear in many other formations based upon the particular needs of the cell and the characteristics of its microtubular network. The mitochondrion is different from most other organelles because it has its own circular DNA (similar to the DNA of prokaryotes) and reproduces independently of the cell in which it is found; an apparent case of endosymbiosis. Scientists hypothesize that millions of years ago small, free-living prokaryotes were engulfed, but not consumed, by larger prokaryotes, perhaps because they were able to resist the digestive enzymes of the host organism. The two organisms developed a symbiotic relationship over time, the larger organism providing the smaller with ample nutrients and the smaller organism providing ATP molecules to the larger one. Eventually, according to this view, the larger organism developed into the eukaryotic cell and the smaller organism into the mitochondrion.

Mitochondrial DNA is localized to the matrix, which also contains a host of enzymes, as well as ribosomes for protein synthesis. Many of the critical metabolic steps of cellular respiration are catalysed by enzymes that are able to diffuse through the mitochondrial matrix. The other proteins involved in respiration, including the enzyme that generates ATP, are embedded within the mitochondrial inner membrane. Infolding of the cristae dramatically increases the surface area available for hosting the enzymes responsible for cellular respiration.

## Methods for Establishing Purity, Intactness, and Functionality of the Fractions

It is often important to determine the degree of purity of the mitochondria and their subfractions. The intactness of the mitochondria is usually established with enzyme assays that take advantage of the fact that an intact membrane is not permeable to certain molecules.

For instance, the outer mitochondrial membrane is impermeable to cytochrome *c*, a 12.5 kDa protein that is too large to pass through the outer membrane pores, and NADH cannot pass through an intact inner membrane.

Functionality is usually tested using an oxygen electrode. If the mitochondria show high rates of oxygen consumption, high respiratory control ratios (>3), and ADP:O ratios close to the theoretical values they are in very good shape. Note that, for mitochondria with an appreciable alternative oxidase activity (measured in the presence of the cytochrome oxidase inhibitor, KCN), the respiratory control ratio and ADP:O ratio should be measured in the presence of an alternative oxidase inhibitor such as salicylhydroxamic acid.

### *Mitochondrial Activities can be Determined* in Situ

It is relatively straightforward to measure the activity of enzymes in the soluble matrix fraction isolated from mitochondria. However, this fraction is very diluted compared to the matrix inside intact mitochondria (300–500 mg protein/ml!) and protein–protein interactions may have been lost. When measuring the activity of enzymes in fractions containing membranes—intact mitochondria and submitochondrial particles—a detergent can often be used to remove the permeability barrier that the inner membrane constitutes.

However, it is also possible to use the ionophore alamethicin to permeabilize the inner membrane to molecules <1000 Da, which makes it possible to measure the activity of matrix and inner membrane enzymes in their native proteinaceous environment. Using alamethicin, it is even possible to measure the activity of cytosolic and mitochondrial enzymes inside intact cells since the ionophore also permeabilizes the plasma membrane.

## The Q-Cycle Explains How Protons Can Be Pumped across the Inner Mitochondrial Membrane

The electron transport chain (ETC) in the mitochondrial inner membrane of most eukaryotes consists of four complexes, complex I to complex IV. Complexes I, III and IV pump protons over the inner mitochondrial membrane when transporting electrons.

For each electron pair (corresponding to one NADH or one oxygen atom), four protons are pumped by complex I, four by complex III, and two by complex IV.

The detailed three-dimensional structures have been determined for the mammalian complexes III and IV, but not for complex I. The structural information has improved the understanding of the electron paths of complexes III and IV, but the proton-pumping function of the latter is still not fully understood. For complex I, neither electron nor proton paths are known in any detail.

Ubiquinone (Q), also called coenzyme Q, is a mobile electron and proton carrier in the inner mitochondrial membrane, structurally and functionally similar to plastoquinone in the thylakoid membrane of plastids (Figure 1). The ability of Q to carry both electrons and protons—and that it can accept and donate them step-wise—is utilized in electron transport and proton pumping (Figure 2) (Nicholls and Ferguson 2002).

To be reduced, it can receive two electrons from complex I or succinate dehydrogenase and take up two protons from the matrix, forming $QH_2$. However, at the oxidizing side (complex III), there is actually both oxidation and reduction taking place. A $QH_2$ will be oxidized to Q at the outer Q-site, releasing two electrons. One of these electrons goes to cytochrome *c*, whereas the other is recycled to reduce an oxidized Q in the inner Q-site—one step to becoming a semiquinone.

The next time a $QH_2$ is oxidized at the outer Q-site, the semiquinone in the inner site will receive a second electron, becoming a $QH_2$. So, from two $QH_2$, we have made one Q and one $QH_2$, which again can be oxidized. In each cycle (i.e., for each electron passing to cytochrome *c*), two protons are taken up from the matrix and two protons released to the intermembrane space. So, the Q-cycle mediates the pumping of an extra proton per electron transported, increasing energy conservation.

If we look at the whole reaction from succinate to cytochrome *c*, we see that for each succinate, four protons are transported from the matrix to the intermembrane space.

However, at the same time, the two electrons are also transported out. So, this corresponds to the net pumping out of two positively charged species. Since, energetically, the mitochondrial membrane potential (which determines the energetic cost of transporting a positive charge out) is much larger than the pH gradient (which determines the cost for electroneutral export of a proton), the complex III proton pumping is less energy-demanding than, for example, complex I, which pumps four protons in their completeness.

## Multiple Energy-Conservation Bypasses in Oxidative Phosphorylation of Plant Mitochondria

The electron transport chain (ETC) in the mitochondrial inner membrane of most eukaryotes consists of four large protein complexes, complex I to complex IV. The electron transport activity of complexes I, III and IV (the latter two, often jointly called the cytochrome pathway) is coupled to extrusion of protons across the inner mitochondria. The electrochemical proton gradient that is formed and maintained in the process is used by the ATP synthase (also called complex V) to make ATP (Figure 1). The plant ETC, however, is highly branched. As a consequence, the plant oxidative phosphorylation system has several alternative pathways of electron transport via type II NAD(P)H dehydrogenases and the alternative oxidase (AOX), and additionally, an alternative proton transporter called the uncoupling protein (UCP).

These proteins mediate bypasses around the proton-translocating multi-protein complexes (I–V): type II NAD(P)H dehydrogenases bypass complex I, AOX bypasses complexes III and IV, and UCP bypasses the ATP synthase (Siedow and Umbach 1995; Vanlerberghe and McIntosh 1997; Møller 2001; Vercesi 2001; Millenaar and Lambers 2003; Fernie et al. 2004; Rasmusson et al. 2004).

The type II NAD(P)H dehydrogenases and the AOX transport electrons without pumping protons, whereas the UCP allows proton flow from the intermembrane space to the matrix without ATP synthesis. The consequence of this is that flux through the energy conservation bypasses will neither contribute to respiratory ATP production, nor will they be controlled by cellular adenylate status. The special features of these energy bypass proteins and discuss the consequences of their existence to the plant.

## *NAD(P)H Dehydrogenases in Plant Mitochondria*

Except for yeast, the ETC of all eukaryotes—including plants—contains a proton-pumping NADH dehydrogenase, complex I, which is inhibited specifically by rotenone. Additionally, plant, fungal and protist mitochondria contain different setups of rotenone-insensitive enzymes capable of oxidizing NADH and/or NAD(P)H and passing the electrons on to ubiquinone.

These NAD(P)H dehydrogenases are classified as type II, whereas complex I belongs to type I. The inner mitochondrial membrane in plants contains such NAD(P)H dehydrogenases, facing either the matrix or the intermembrane space, thus oxidizing NAD(P)H either from the matrix or the cytoplasm.

The outer membrane of all mitochondria contains NADH-ferricyanide and NADH-cytochrome *b* reductase activities that are insensitive to rotenone and to the complex III inhibitor, antimycin A.

## *There are Three NAD(P)H Dehydrogenases on the Inner Matrix Surface of the Inner Membrane*

Much of the information about the number of matrix-facing NAD(P)H dehydrogenases has come from experiments using inside-out inner membrane vesicles. Such vesicles have the matrix surface facing the medium and the external NAD(P)H dehydrogenases hidden inside the vesicles and unable to interfere with the assays. Recently, this method has been complemented with a technique where one uses a peptide called alamethicin to make a channel for the substrate through the inner membrane. Experiments with these methods show the presence of three NAD(P)H dehydrogenase activities on the matrix side of the inner membrane:

- Complex I pumps protons, is $Ca^{2+}$-independent and is inhibited by rotenone and diphenylene iodonium (DPI). Complex I has a high affinity for NADH.
- $ND_{in}$(NADH) does not pump protons, is $Ca^{2+}$-independent and DPI insensitive. The enzyme has a tenfold lower affinity for NADH than complex I.
- $ND_{in}$(NADPH) does not pump protons, is $Ca^{2+}$-dependent and very sensitive to DPI.

Because of the contrasting affinities of complex I and $ND_{in}$(NADH) for NADH, only complex I is engaged at low matrix concentrations of NADH. However, when the NADH concentration in the matrix increases (e.g., if the ETC is restricted by lack of ADP in the cell), $ND_{in}$(NADH) can become engaged. When electrons are transported through the type II NAD(P)H dehydrogenases, only the two proton pumping sites of the cytochrome pathway are involved in the flow of electrons from matrix NADH to oxygen. As a consequence, the theoretical ADP/O yield is 1.5 unless the AOX or the UCP are active, in which case it will be even lower.

### *The External NAD(P)H Dehydrogenases are $Ca^{2+}$-Dependent*

Plant mitochondria can oxidize cytosolic NADH and NAD(P)H directly via external dehydrogenases. The NADH and NAD(P)H dehydrogenase activities are catalyzed by separate enzymes. None of the external dehydrogenases pumps protons, which means that, at most, two sites of proton-pumping are involved when the electrons move from the dehydrogenases to oxygen. So, as for $ND_{in}$(NADH), the theoretical maximum ADP/O yield is 1.5.

The oxidation of exogenous NADH or NAD(P)H by most plant mitochondria requires $Ca^{2+}$, and as a result, is inhibited by EDTA and EGTA, chelators of di-and trivalent cations. *In vitro*, the $Ca^{2+}$concentration needed for half-maximal activity of NADH oxidation is 0.1–0.5 $\mu$M, depending on ionic environment and respiratory state This is in the same range as the 0.1–0.2 $\mu$M free $Ca^{2+}$ thought to be present in the cytosol of a resting plant cell. Therefore, the external NAD(P)H oxidation may become activated by the increased $Ca^{2+}$-concentration associated with cellular excitation due to external conditions.

Mammalian mitochondria lack external NAD(P)H dehydrogenases. Instead, they have a $Ca^{2+}$-dependent glycerol-3-phosphate dehydrogenase on the outer surface of the inner membrane, which donates electrons directly to the ubiquinone pool.

The dihydroxyacetone phosphate formed is converted back into glycerol-3-phosphate by a cytosolic NAD-dependent glycerol-3-phosphate dehydrogenase. In this way, the two low-molecular-weight compounds, glycerol-3-phosphate and dihydroxyacetone phosphate, shuttle redox equivalents from the cytosol, across the outer membrane and to the inner membrane to yield the same amount of ATP as is produced by the plant mechanism. Yeast have both external NADH dehydrogenases and a glycerol-3-phosphate shuttle, and since a mitochondrial glycerol-3-phosphate dehydrogenase has been found in plant mitochondria, plants may also have both systems.

Another way to oxidize cytosolic NAD(P)H that is found in mammalian mitochondria is to use the malate/aspartate shuttle to tranfer reductant (a collective term for high-energy electrons carried by, for example, NAD(P)H, ferredoxin or an organic acid redox couple like malate/oxaloacetate) into the matrix as NADH that can be oxidized by the internal NAD(P)H dehydrogenases. However, it is not clear whether this shuttle can work for import of cytosolic reductant into plant mitochondria.

In the opposite orientation, NADH is shuttled from the matrix to the cytosol via the malate/oxaloacetate shuttle. In this way, the external NADH dehydrogenase can also work as a bypass around complex I. Support for this bypass has been seen in the form of an upregulation of external NADH dehydrogenase activity in mutants where complex I or an internal type II NAD(P)H dehydrogenase has been inactivated.

### *The Molecular Identities of the Rotenone-Insensitive NAD(P)H Dehydrogenases*

The type II NADH dehydrogenases from *E. coli* and yeast are small, single polypeptide proteins of around 50 kDa. They oxidize only NADH and contain FAD as an electron-accepting prosthetic group.

Attempts to purify the type II NAD(P)H dehydrogenases have been unsuccessful in identifying the plant proteins. However, the characterization of potato homologues to the type II NADH dehydrogenases in *E. coli* and yeasts have provided candidate plant proteins to try and assign to the different reported activities. One protein, NDA1, is internal and the other, NDB1, external of the inner mitochondrial membrane. The latter was shown to be an external calcium-dependent NAD(P)H dehydrogenase. Like a related fungal NAD(P)H dehydrogenase, NDB1 contains a motif for Ca-binding, indicating that in the activation of plant and fungal NAD(P)H dehydrogenase by $Ca^{2+}$, the ion binds directly to the enzyme.

In Arabidopsis, a weed for which the nuclear genome sequence is known, small families of both NDA-type and NDB-type proteins are encoded by nuclear genes. One of the NDA-type proteins has been shown to be an internal NAD(P)H dehydrogenase. Additionally, a third type of type II NAD(P)H dehydrogenase was found in Arabidopsis, the NDC1. This protein is most closely related to cyanobacterial enzymes, suggesting that the gene entered the eukaryotic domain with the plastid, later to be transferred to the nucleus.

### *The Alternative Oxidase*

The presence of the AOX is one of the features that sets plant (and fungal) mitochondria apart from mammalian mitochondria. This enzyme has therefore received a lot of attention from plant scientists and we now have a fairly good understanding of the regulation of its structure and activity.

The AOX is a quinol-oxygen oxidoreductase and it does not pump protons. The oxygenase transfers electrons from ubiquinol (the reduced form of ubiquinone) to oxygen and generates water as the end-product of the reaction, so four electrons must be transferred to oxygen.

The precise mechanism is not known, but the protein has been shown to contain a catalytical di-iron centre. The functional form of the enzyme is a dimer with the two polypeptides either covalently or non-covalently bound to each other (Figure 2). The predicted structure places the protein embedded into the inner leaflet of the inner mitochondrial membrane, with the active site and the regulatory cysteines exposed to the matrix.

When electrons from NADH oxidation flow through the AOX, at least two of the three sites of proton translocation are bypassed, so energy conservation in the form of ATP is much smaller when the AOX is active. The activity of the AOX must therefore be carefully regulated. There are often several AOX isoforms in each plant species—for example, five genes encoding AOX proteins are present in Arabidopsis. The relative amounts of three isoforms of AOX vary with development, and cellular redox status, as well as external factors.

The AOX is present in the inner membrane as a dimer in which the two subunits can be covalently linked through a disulfide bridge between cysteine residues. The reduced non-covalently linked dimer is the active form, and this form appears to be the predominant one *in vivo*. It was recently shown that a mitochondrial thioredoxin system regulates AOX. Thioredoxins regulate enzymes by reducing disulfide bridges, and it is likely that the regulation involves the redox-active cysteines.

The presence of thioredoxin regulation connects the AOX activation level to the mitochondrial NAD(P)H pool and the oxidation of citric acid cycle metabolites by NADP-utilizing enzymes. The activity of AOX is also stimulated by 2-oxo organic acids—in particular, pyruvate, which activates by interacting with the same cysteine that is involved in dimerization. This activation may be a feed-forward mechanism that upregulates the capacity of the ETC when the glycolytic end-product is available. Both the cytochrome pathway and the AOX pathway use ubiquinol as the substrate. In the absence of pyruvate, the cytochrome pathway is saturated at 40% $Q_{red}/Q_{tot}$ (40% of the UQ is reduced), at which point the AOX pathway is not engaged at all. The AOX pathway starts accepting electrons at 50% reduction and is not fully saturated at 80%.

In the presence of pyruvate, the AOX becomes engaged at a much lower $Q_{red}/Q_{tot}$ (20%), whereas the cytochrome pathway is unaffected. It is puzzling, however, that the normal *in vivo* concentration of pyruvate appears to be higher than the concentration of pyruvate required to give maximal stimulation of AOX activity *in vitro*. It thus appears that there is little regulation of AOX activity via pyruvate under physiological conditions.

### *AOX Activity Can Be Measured In Vivo*

For many years, benzhydroxamic acid derivatives,, were used to quantify the capacity and activity of the AOX pathway in both isolated mitochondria and intact tissues. AOX capacity was determined from the rate of respiration in the presence of KCN (which blocks the cytochrome pathway), and corrected for residual respiration in the presence of both KCN and SHAM.

AOX *in vivo* activity was determined from the decrease in the rate of respiration when SHAM was added in the absence of KCN. The assumption was that the cytochrome pathway was always used first and that the AOX would be engaged only when the cytochrome pathway was saturated. However, as we have seen above, this assumption is not correct. The consequence of this is that *in vivo* activity of AOX cannot be measured using inhibitors alone.

A better method for measuring AOX *in vivo* activity is now available. It makes use of the fact that cytochrome *c* oxidase and AOX discriminates differently between $^{18}O$ relative to $^{16}O$.

The so-called oxygen discrimination of the sample (tissue or mitochondria) can be measured by mass spectrometry: (1) with KCN to resolve the oxygen discrimination by AOX; (2) with SHAM to resolve the oxygen discrimination by cytochrome *c* oxidase; and (3) without inhibitors to resolve the discrimination value resulting from the activity of both enzymes.

From these values, the percentage contribution by AOX and cytochrome *c* oxidase to the total uninhibited rate of respiration can be calculated. An example of the use of the oxygen discrimination technique with an intact tissue is given below. A remaining problem in the estimation of AOX activity *in vivo* is that measurements can only be made in darkness (to avoid interference from photosynthetic oxygen exchange).

### Plant Mitochondria also Contain an Uncoupling Protein

Mammalian mitochondria contain neither type II NAD(P)H dehydrogenases nor AOX. However, they have a UCP that increases the proton permeability of the inner mitochondrial membrane and, in that way, dissipates the proton gradient. This is another mechanism for reducing the ATP production and increasing heat production, for example, for thermoregulation.

Surprisingly, plant mitochondria also contain a UCP. We do not know for certain why plant mitochondria should require two different mechanisms for achieving the same end result. The UCP is related to anion carriers.

Its *in vitro* transport activity is dependent on free fatty acids and is inhibited by nucleotides like ATP and GTP but stimulated by superoxide. The latter result indicates that the UCP may protect against oxidative stress.

### The Physiological Impact of Energy Bypass Proteins

The respiratory rates of some thermogenic flowers are the highest among plants and, in fact, exceed even those of warm-blooded animals. The external NADH dehydrogenase, AOX and UCP appear particularly active in thermogenic flowers where they are primarily responsible for heat production. At the same time it is clear that the respiratory rate of most plant tissues is insufficient to generate an appreciable increase in tissue temperature.

AOX is up-regulated by cold stress, although this is clearly not to heat the tissue. However, during cold stress, reactive oxygen species (ROS) are formed, and also other stress treatments that lead to elevated ROS will induce AOX. ROS (e.g., superoxide and hydrogen peroxide) are formed as by-products of electron transport under aerobic conditions.

ROS can cause damage to proteins, lipids, and DNA and the cell must therefore limit their formation. AOX helps prevent overreduction of the ETC and thus lowers ROS production. Supporting this, Maxwell et al. (1999) demonstrated the importance of the AOX for limiting ROS formation in living plant cells.

Overexpression of AOX lowered the steady state level of ROS, whereas suppression increased it. This result also points to the mitochondrion as an important site of ROS generation in plant cells. With the recent discovery of ROS stimulation of plant UCP and the increased ROS resistance in plants overexpressing UCP, the picture has become more complex.

By decreasing the resistance for electron transport through the proton-pumping complexes, UCP activity may, like AOX, lead to decreased overreduction and ROS formation. Another factor speaking in favour of UCP having a role in ROS protection is that the AOX is inhibited by an ROS product.

Both type II NAD(P)H dehydrogenases and AOX may act as reductant overflow enzymes by reoxidizing surplus NADH to oxygen. This can prevent inhibition of the citric acid cycle under conditions in which the ADP concentration is low, and thus the proton-pumping complexes are restricted by a high membrane potential. In this way, the citric acid cycle can supply carbon skeletons for biosynthesis, even if the ATP level in the cell is high. The concept of overflow is also relevant for photorespiratory conditions, under which there is a massive flow of glycine that has to be oxidized by the mitochondria. As much as half of the NADH produced in this process is probably exported, but the other half has to be oxidized promptly. There are several reasons to believe that a reductant overflow via energy bypass proteins is involved.

The $ND_{in}$(NADH) and AOX capacity in leaf mitochondria is higher than in mitochondria from heterotrophic tissue, and both proteins are down-regulated by dark treatment. Consistent with these observations, AOX activity *in vivo* is up-regulated by light (Ribas-Carbo et al. 2000), AOX capacity is induced by greening of etiolated tissue, and expression of the *nda*1 gene [encoding an internal NAD(P)H dehydrogenase] in potato and Arabidopsis is light-dependent.

Mutant tobacco plants, lacking the proton-pumping complex I can still survive, apparently aided by the action of $ND_{in}$(NADH), which oxidizes matrix NADH with a lower ATP yield.

These plants, however, show decreased photosynthetic rates, indicating that the $ND_{in}$(NADH) cannot fully compensate for the missing complex I. Also, external NAD(P)H dehydrogenases have been suggested to support photosynthesis. These enzymes may reoxidise NAD(P)H that is shuttled out of the chloroplast in order to balance the NAD(P)H to ATP ratio in the stroma. In summary, there are many indications that mitochondria—especially the energy bypass proteins—have roles in supporting photosynthesis, for example by oxidizing excess NAD(P)H.

The AOX pathway is involved in the acidification/deacidification cycle of CAM plants. CAM plants fix $CO_2$ in the form of malate during the night. This is done by cytosolic PEP carboxylase, and the malate produced is stored in the vacuole.

The $CO_2$ fixed is released during the daytime by malic enzymes, located in cytosol and mitochondria. Robinson et al. (1992) used the oxygen discrimination technique to measure the contribution of the AOX pathway to whole-leaf respiration in the CAM-plant, *Kalanchoë daigremontiana.*

They found that the respiratory rate of the leaves increased by 35% in the early hours of the day. This increase was due to increased AOX activity. A more detailed time course showed that the AOX activity increased two-to threefold during the early day at the same time as the malate was metabolized. As soon as the malate had disappeared by mid-day, the AOX activity returned to the low level observed during the night. A reasonable conclusion is that a substantial amount of the malate from the vacuole is oxidized via malic enzyme in the mitochondria. This generates NADH in the matrix, which is oxidized to a large extent via the AOX. The released $CO_2$ can then be refixed by the Calvin cycle in the chloroplast.

## $F_oF_1$-ATP Synthases: The Worlds Smallest Rotary Motors

$F_oF_1$-ATP synthases (also called F-ATPases) are present in the inner membranes of mitochondria, chloroplasts, and bacteria. These large, multisubunit enzymes consist of a water-soluble catalytic complex ($F_1$) attached to an integral membrane protein complex ($F_o$) that transports protons across the membrane. The $F_1$ complex is composed of at least five different types of subunits (three $\alpha$, three $\beta$, one $\gamma$, one $\delta$, and one å).

When the $F_1$ complex is dissociated from the membrane, it is active as an ATPase. In fact, under the appropriate conditions the intact $F_oF_1$-ATP synthase can run in reverse and act as a proton pump, using the energy of ATP hydrolysis to move $H^+$ across the membrane. Our understanding of how ATP is synthesized was advanced by Paul Boyer at the University of California, Los Angeles, who proposed the binding-change mechanism for catalysis by F-ATPases. From the detailed understanding of how the ATPases function, he proposed that the binding-change mechanism for ATP synthesis contains three important components:

1. The major energy-requiring step is not the synthesis of ATP from ADP and $P_i$, but the release of ATP from the enzyme.
2. Substrate is bound and products are released at three separate but interacting catalytic sites, corresponding to the three catalytic subunits (b subunits). Each catalytic site can exist in one of three conformations: tight, loose, or open.
3. The binding changes are coupled to proton transport by rotation of the g subunit. That is, the flow of protons

down their electrochemical gradient through the $F_0$ complex causes the g subunit to rotate.

Rotation of the γ subunit then brings about the conformational changes in the catalytic complex that allow the release of ATP from the enzyme, and the reaction is driven forward. The reverse occurs when the enzyme functions as an ATP-driven proton pump.

The first two predictions of the binding-change model are supported by many lines of evidence, mainly kinetic studies, and are now generally accepted. However, the prediction of a rotary mechanism for coupling proton flow to ATP synthesis has been more difficult to demonstrate. Two major breakthroughs have led to confirmation of the third prediction, as well.

The first breakthrough was the determination of the crystal structure of the $F_1$-part of bovine mitochondrial ATPase by the laboratory of John Walker in Cambridge, England. The crystal structure showed that the three catalytic β subunits differ in their conformations and in the nucleotide bound to them, consistent with the binding-change mechanism.

Even more exciting was the discovery that the γ subunit is inserted like a shaft through the centre of the catalytic complex, which consists of three α subunits and three β subunits arranged alternately in a doughnutlike structure. Moreover, the interface between the γ subunit and the α and β subunits is highly hydrophobic.

The hydrophobicity of the interface minimizes the interactions between the subunits, consistent with the rotation of the γ subunit within the hole formed by the catalytic complex. In other words, the γ subunit looks like a molecular bearing lubricated by a hydrophobic interface. Although many questions were answered by elucidation of the crystal structure of the $F_1$-ATP synthase, the rotational model cannot be tested by a static "snapshot" of the enzyme. Definitive demonstration of rotation requires a video recording of the spinning of the enzyme in real time. But although they are large for proteins, $F_1$-ATP synthases are still far too small to be visualized in a light microscope.

To visualize the rotation of the enzyme, Masasuke Yoshida and his colleagues at the Tokyo Institute of Technology came up with an ingenious method to make the enzyme much larger than it is. As Figure below shows, they attached an actin filament labelled with a fluorescent dye to the base of the γ subunit using another protein as a "glue." They then attached the $F_1$ complex upside down to a glass surface.

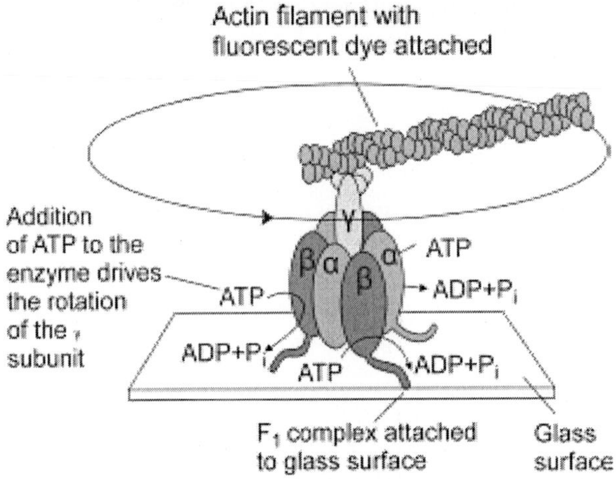


*Figure : A method for visualizing rotation of the g subunit. A fluorescently labelled actin filament was attached to one protruding end of the g subunit. The $F_1$ complex was then attached upside down to a coverslip. When ATP was added to the coverslip, the actin filament rotated. (After Noji et al. 1997.)*

If the γ subunit rotates with respect to the catalytic complex, the actin filament should swing around with it. Since the filament is very long compared to the ATP synthase (about 1 μm), its rotation should be visible in a fluorescence microscope. In other words, the fluorescently tagged actin filament, which is large enough to visualize in a light microscope, reports the rotation of the γ subunit. The results were spectacular! When ATP was added to the modified enzyme, the actin filaments were seen to swing around in a circle at as much as 4 revolutions per second in a fluorescence microscope. To give some idea of scale, if you were a γ subunit, this rotation rate would be equivalent to swinging a several hundred meter-long rod around your head at 4 revolutions per second in water! However, the measured velocity is undoubtedly an underestimation of the actual velocity in vivo, because of the enormous torque required to swing such a large mass. Demonstration of the rotary motion of the $\gamma$ subunit made it possible to put together a model of how the ATP synthase works. For their contributions to elucidation of the mechanism of ATP synthesis, Paul Boyer and John Walker shared half the Nobel prize in physiology or medicine in 1997. The other half went to Jens Skou for his pioneering work on the $K^+$,$Na^+$-ATP synthase, the mammalian counterpart of the plant plasma membrane $H^+$-ATP synthase.

## Transport Into and Out of Plant Mitochondria

Plant mitochondria must exchange metabolites and information (signal transduction) with the cytosol. To give some central examples, the end products of glycolysis, pyruvate and malate, need to be taken up, and citric acid cycle intermediates exported to be used as building blocks in biosynthetic processes.

Likewise, ADP and $P_i$ must be taken up and ATP exported for the mitochondrion to fulfil its role as the powerhouse of the cell. To carry out the transport of this large number of molecules, plant mitochondria possess a family of mitochondrial carrier proteins. These are transmembrane proteins with a common general structure and they typically transport relatively small charged molecules, like metabolites. In Arabidopsis there are almost 60 members, most of which are not functionally characterized, giving a hint to the complexity of the mitochondrial transport processes. However, mitochondria exchange more than metabolites, and therefore also other types of transporters are involved. The outer membrane is permeable to molecules of less than 10 kDa, (i.e., all non-polymeric compounds). The permeability results from the presence of pores, which appear to be open in most *in vitro* experiments. Recent studies in animals have characterized pore-forming proteins in the mitochondrial outer membrane, called porins, which appear to regulate transport across the membrane. Plant mitochondria also have porins but their characterization awaits further research. In our discussion here we will focus on the regulation of mitochondrial transport across the inner mitochondrial membrane. We will first consider the relatively few molecules that can move across the inner membrane without the help of a specific carrier protein. We will then discuss the various metabolite transporters and, finally, briefly look at the transport of proteins and RNA.

### *Small, Neutral, and Hydrophobic Molecules Diffuse across the Inner Membrane*

Irrespective of size, molecules that either are charged or are very hydrophilic can only cross a lipid bilayer with the help of a specific protein carrier.

Small, *neutral* molecules can diffuse across the lipid bilayer and do not require protein transporters. Among those are $O_2$ and $CO_2$, two important molecules for respiration. $O_2$ is needed on the inner surface of the inner membrane where the active sites of both cytochrome *c* oxidase and the alternative oxidase are located. The $CO_2$, produced as an end product in the citric acid cycle, needs to diffuse out of the mitochondrion, ultimately to be lost to the surroundings or re-fixed in the chloroplasts. In addition, however, it is possible that $CO_2$ can also be transported out of the mitochondrion as bicarbonate ($HCO_3^-$). Mitochondria are osmotically active and swell in a hypotonic medium. This means that the inner membrane is also permeable to $H_2O$, but it is not clear whether this occurs by diffusion or if aquaporins may be involved.

For technical reasons, it is of importance that several well-known inhibitors of mitochondrial function can diffuse into mitochondria because they are uncharged and hydrophobic. These are rotenone, the specific inhibitor of complex I, antimycin, the specific inhibitor of complex III, HCN, the protonated form of the cyanide ($CN^-$) inhibitor of complex IV (and other heme-containing enzymes), and oligomycin, the specific inhibitor of the $F_oF_1$-ATP synthase.

### *Transport of Inorganic Ions*

Inorganic phosphate ($H_2PO_4^-$) is taken up into the matrix in exchange for a hydroxyl ion ($HO^-$). The energy for the uptake is supplied by the proton gradient (–pH). We will see below that the phosphate gradient, in turn, drives transport of other metabolites. Several other inorganic ions—$Mg^{2+}$, $Ca^{2+}$, $K^+$, and $NH_4^+$—are involved in mitochondrial metabolism, but the mechanism by which they pass the inner membrane of plant mitochondria is not well understood. In mammalian mitochondria, uptake of $K^+$ through a $K^+$-specific channel, and extrusion via a $K^+/H^+$ antiporter, is involved in matrix volume regulation and cellular signalling (Garlid and Paucek 2003). A similar system of $K^+$ transport is present in plant mitochondria. Both mammalian and plant mitochondria contain an uncoupling protein (UCP). This protein facilitates the movement of protons across the inner membrane and therefore uncouples electron transport and decreases ATP yield.

### *Transport of Adenylates*

One of the important functions of mitochondria is to supply the cell with ATP. To do this, mitochondria have an adenine nucleotide transporter that facilitates the exchange of $ADP^{3-}$ (inward) for $ATP^{4-}$ (outward). This exchange is driven by the membrane potential (inside negative), since there is a net outward movement of one negative charge. In combination with the phosphate carrier reaction, this means that the uptake of one ADP and one phosphate and the export of one ATP (i.e., the transport event connected to mitochondrial ATP synthesis) costs one vectorial proton—the electrical component for adenylate exchange and the "pH component for phosphate uptake.

### *Transport of Citric Acid Cycle Intermediates and Related Compounds*

Pyruvate and malate, the end products of glycolysis, both have transporters in the inner membrane. Pyruvate is taken up in exchange for a hydroxyl ion, and $malate^{2-}$ in exchange for $H_2PO_4^{2-}$. The transporter for malate also takes $succinate^{2-}$ and is therefore called the dicarboxylate transporter. $Malate^{2-}$ or $succinate^{2-}$ can, in turn, exchange for $citrate^{2-}$ on the tricarboxylate transporter. $Malate^{2-}$ can also exchange for 2-$oxoglutarate^{2-}$ via a special transporter. An interesting transporter, the oxaloacetate transporter, exchanges $oxaloacetate^{2-}$ for $malate^{2-}$ (or $citrate^{2-}$), which permits the shuttling of reducing equivalents. The joint activity of these transporters allows plant mitochondria to import or export most citric acid intermediates. Finally, the glutamate/aspartate transporter facilitates the uptake of glutamate. In mammalian mitochondria, glutamate is taken up together with a proton, and it is thus driven by both the electrical and the chemical proton gradient. A similar transporter is present in plant mitochondria, but a proton co-transport has not been shown, so the transport may be passive in plants. Recently, several additional transporters for metabolites have been identified, and it is likely that certain setups of transporters are only expressed under special developmental conditions, like the mobilization of storage nutrients in developing seedlings. In several cases the involvement of the mitochondrion in cellular metabolism implies that certain metabolites are taken up across the inner membrane, but the transporter has not yet been described. This is the case for the massive flux of glycine into—and $CO_2$, serine and $NH_4^+$ out of—mitochondria in $C_3$ leaves during photorespiration This is also the case for proline, which accumulates to high concentrations in the cytosol during episodes of plant water stress, and it is then broken down during the rehydration phase by a pathway involving two mitochondrial enzymes.

### *Transport of Coenzymes*

Coenzymes are either synthesized inside the mitochondrion or imported. There is some evidence that, in addition to ATP, at least $NAD^+$, thiamine pyrophosphate, and coenzyme A are imported across the inner membrane. The final step in ascorbic acid biosynthesis has recently been shown to be on the outer surface of the inner mitochondrial membrane. In view of the accumulating evidence that ascorbate has an important function in removing ROS inside the mitochondrion, we should expect to find an ascorbate transporter in the inner mitochondrial membrane. The situation with folate (involved in 1C transfers) is the opposite to that with ascorbate: The only site of synthesis of many of the folate metabolites in the plant cell is the mitochondrial matrix. We can therefore expect that these molecules are exported via specific transporters.

### *Transport of Reducing Equivalents—Metabolic Shuttles*

Mammalian mitochondria cannot oxidize cytosolic NADH or NAD(P)H directly, so they make use of "shuttle mechanisms." An exchange of metabolites by inner membrane transporters takes place through these shuttles so that a reduced compound moves into the matrix in exchange for a more oxidized compound moving out into the cytosol. In this way, reducing equivalents are transported into the matrix ultimately to be oxidized by the respiratory chain. The shuttles can also be used to transport reducing equivalents in the opposite direction. Although plant mitochondria can oxidize cytosolic NADH and NAD(P)H directly by the two external NAD(P)H dehydrogenases, they can also use these two metabolic shuttles. The malate/oxaloacetate shuttle uses malate dehydrogenase in the cytosol and in the matrix to catalyze the interconversion of malate (reduced) and oxaloacetate (oxidized).

These two compounds are exchanged on the oxaloacetate transporter. This exchange, however, is not driven by the electrochemical proton gradient across the inner membrane, so it can only move reducing equivalents from a relatively reduced compartment to a relatively oxidized compartment. Since the mitochondrial matrix is more reduced than the cytosol, the oxaloacetate transporter is likely to work primarily in the export of reducing equivalents.

Under photorespiratory conditions, large amounts of glycine are oxidized in the mitochondria to give NADH while, at the same time, NADH is required in the peroxisomes to reduce hydroxypyruvate to glycerate. It is thought that as much as 50% of the matrix NADH is exported, probably via the malate/ oxaloacetate shuttle.

The more complex malate/aspartate shuttle involves two matrix enzymes and two cytosolic enzymes as well as two inner membrane transporters. In mammals, it is powered by the electrochemical proton gradient in the direction of the import of reducing equivalents. However, since the glutamate/aspartate exchanger may be passive in plants, this shuttle in plants may similarly be independent of the electrochemical proton gradient. If so, it would (like the malate/oxaloacetate shuttle described above) only be able to function as an exporter of mitochondrial reductant.

### *Transport of Proteins—Protein Import*

Plant mitochondria contain their own DNA (mtDNA) and ribosomes and are capable of synthesizing the proteins encoded in the mtDNA. However, plant mtDNA encodes only a few of the proteins found in the mitochondrion. The great majority of the mitochondrial proteins have to be imported from the cytosol where they are synthesized on free ribosomes. These are synthesized with a presequence or targeting peptide in the N-terminal end, which is recognized by receptors on the outer mitochondrial membrane. The import of proteins requires a complex machinery with two multisubunit protein complexes—one in the outer membrane (TOM, transporter of the outer membrane) and one in the inner membrane (TIM, transporter of the inner membrane). Concomitant with, or immediately afterwards the import of the protein into the correct mitochondrial compartment, the targeting peptide is often cleaved off to yield the mature and active protein.

### Transport of RNA

Mammalian mtDNA encodes the full complement of tRNA required for mitochondrial protein synthesis. In contrast, plant mtDNA only encodes some of the tRNAs required for protein synthesis within the mitochondrion. The missing tRNA must be imported by a yet-to-be-elucidated import mechanism.

## The Genetic System in Plant Mitochondria has Some Special Features

As in plastids, the genetic system in mitochondria differs both structurally and functionally from the nuclear genome and its gene expression. On the other hand, mitochondria share several genetic features with bacteria.

For example, several genes are transcribed together and the ribosomes are of a prokaryotic type. This is in line with the evolutionary origin of the mitochondrion as an endosymbiont of the α-subclass of Proteobacteria, a group including the intracellular pathogen *Ricketsia prowazekii.*

Plants are also distinctly different from other eukaryotes in their mitochondrial DNA (mtDNA) and its expression. For example, the plant mtDNA is very large and highly variable in size between species.

The presence of several repeats enables the genome to recombine, changing the gene order or even causing a split into smaller subgenomic molecules, each carrying a subset of the genes. Though the size of mtDNA varies between plants, the number of genes appears to be similar in all plants. In the fully sequenced 367 kilobasepair (kb) mtDNA of *Arabidopsis thaliana,* the known genes (constituting 65 kb) are scattered in a background of noncoding sequence. A similar situation can be found in the sequenced mtDNA of sugarbeet.

Plant mtDNA is quite different from the much smaller (16 kb) single mtDNA molecule of mammals in which every base pair has a function in coding for protein or mediating transcription and replication, and neither introns nor repeats exist. The expression of the mitochondrial genes in plants is carried out by RNA polymerases that are unlike the nuclear polymerases, but related to the enzymes in bacteriophages.

For several mitochondrial genes, these phage-type polymerases start up transcription from multiple promoters. Many genes in plant mtDNA are broken up into exons by introns. The introns are removed in the processing of the primary RNA into a mature mRNA that can be translated into the correct polypeptide.

The introns in mitochondria belong to the so-called self-splicing type II group, in which splicing is defined by the intron structure.

The splicing can also work in the *trans* mode, in which it can join segments from separate RNA molecules, and thus assemble protein-coding reading frames even if the exons are distributed on different primary transcripts. This allows individual exons of a gene to reside independently in the genome, even in different subgenomic DNA molecules.

Because mitochondrial genes are often co-transcribed with unrelated genes into a polycistronic primary transcript, RNA processing also includes endonucleolytic reactions, so that each mature mRNA encodes only one protein. Like mRNAs for nuclear encoded genes, mitochondrial mRNAs are polyadenylated. However, instead of stabilizing (the case for nuclear encoded mRNAs), the mitochondrial mRNA polyadenylation mediates degradation. Instead, stem-loops stabilise the mRNA in plant mitochondria.

In plant mitochondria, an additional RNA processing event takes places—RNA editing. In this process, certain cytidines (C) are deaminated into uridines (U). Rarely, U is aminated to C instead.

This means that the coding sequence is changed at the RNA level, in order to restore the correct information for translation where the DNA has diverged. Editing sites mainly occur in DNA encoding protein sequences and functional RNA, and are rare in noncoding mtDNA. In principle, the higher the importance of an individual base, the higher the probability that it constitutes an editing site. Despite the considerable amount of bases that can be edited (up to 15% of the bases of a gene), the functional relevance of this process and the consequent divergence in mtDNA sequence is not understood. Mitochondrial RNA editing has been found in land plants, but not in algae. It also occurs, though to a lesser extent, in plastids.

The mitochondrial genome encodes only a small proportion of the mitochondrial proteins; less than 40 as compared to up to thousands of nuclear-encoded mitochondrial proteins. Through evolution, genes have been transferred from the mitochondrion to the nucleus. This is a process that still takes place, at least in plants. For example, subunit 2 of cytochrome oxidase is encoded by the mitochondrial genome in all investigated eukaryotes except for some legumes and protists. Within the legume family, several intermediates of the gene transfer have been found. For example, a functional gene can be found in the mitochondrial genome and an unexpressed one in the nuclear genome, or vise versa.

In order for a mitochondrial gene to be successfully transferred to the nucleus, the protein-coding frame must be integrated into the chromosome, and several other modifications must be made. For example: (1) a promoter for nuclear transcription must be inserted at the beginning of the gene, (2) the mRNA should acquire sequences specifying the addition of a poly-A tail, and (3) a targeting sequence is necessary for the cytosolically synthesized protein to be imported into the mitochondrion. For proper function, elements regulating the expression in different cell types and under different conditions must also be included. Since the coding sequence for most mitochondrial genes in plants is edited at the RNA level, the mitochondrial genomic sequence will not code for a functional protein if inserted into the nucleus. Therefore, the DNA that is integrated into the nuclear genome must in most cases be made by reverse transcription from mitochondrial mRNA. Recently, upon sequencing the nuclear genome of *Arabidopsis thaliana*, a copy of most of the mtDNA was found integrated in a nuclear chromosome, indicating that large segments of DNA can move between cellular compartments.

However, most of these mitochondrial genes transferred to the nuclear genome cannot produce functional proteins unless bases corresponding to editing sites are corrected by mutation.

However, this is unlikely to take place before other mutations have further disrupted the sequence. As opposed to other mitochondrial genomes, plant mitochondria utilize the universal genetic code. Thus, this factor will not prevent or delay the transfer of genes between plant organelles as it may do in other eukaryotes.

The special features of the mitochondrial genetic system in plants also have consequences for the study of mitochondrial genes. Due to RNA editing and *trans*-splicing, a reading frame is more difficult to identify based only on genomic sequence. Also, protein sequences in databases are automatically translated from genomic DNA sequences without compensating for RNA editing in many cases. This protocol would not work for mitochondrial genes.

## Does Respiration Reduce Crop Yields?

Plant respiration can consume an appreciable amount of the carbon fixed each day during photosynthesis over and above the losses due to photorespiration. To what extent can changes in a plants respiratory metabolism affect crop yields?

Attempts to establish a quantitative relationship between respiratory energy metabolism and the various processes going on in the cell have led to a break-up of respiration into two components. In growth respiration, reduced carbon is processed to bring about the addition of new biomass.

The other component, maintenance respiration, is needed to keep existing, mature cells in a viable state. Utilization of energy by maintenance respiration is not well understood, but estimates indicate that it can represent more than 50% of the total respiratory flux.

Although numerous questions remain regarding these issues, there are several empirical examples of relations between plant respiration rates and crop yield. In the forage crop, perennial ryegrass (*Lolium perenne*), yield increases of 10% to 20% were correlated with a 20% decrease in the leaf respiration rate.

Similar correlations have been found for other plants, including corn (*Zea mays*) and tall fescue (*Festaca arundinacea*). However, later investigations have shown that "selection for low rates of mature leaf respiration is not an appropriate method to select for high-yielding cultivars in perennial grasses".

Although there is a potential for increasing crop yield through reduction of respiration rates, a better understanding of the sites and mechanisms that control plant respiration is needed before such changes can be exploited commercially in a systematic fashion by plant physiologists, geneticists, and molecular biologists. Furthermore, much remains to be learned about the general applicability of such observations and the conditions under which slower-respiring lines could be at a disadvantage, causing a reduction in crop yields rather than an increase. Two recent studies illustrate the present difficulty of predicting the effect of directed changes at the molecular level on plant productivity:

Nunes-Nesi et al. (2005) found that the productivity of transgenic tomato plants with a reduced activity of mitochondrial malate dehydrogenase was increased as compared to the wild type. Although the respiratory activity of mitochondria isolated from the transgenic plants was unchanged or higher than in the wild type, the rate of leaf respiration was reduced in the transgenic plants. Photosynthesis was markedly increased in the transgenic lines, possibly linked to increased levels of ascorbate.

Sieger et al. (2005) studied the effect of the alternative oxidase on the growth of tobacco cell cultures. Considering the energy-wasteful nature of the alternative oxidase one might expect that a cell line lacking the enzyme would grow faster than the wild type. Secondly, alternative oxidase has a role in the response of plants to oxidative stress, so a cell line lacking the alternative oxidase would also be expected to handle abiotic stress less well.

The results were surprising—the transgenic cell line with a very low expression of alternative oxidase grew as fast as the wild-type cells under normal, nutrient-sufficient conditions, and faster than the wild type under conditions of macronutrient limitation (low phosphate or low nitrogen).

It appears that the alternative oxidase is an important factor in modulating the growth rate in response to nutrient availability. Plant respiration involves an intricate metabolic network of interacting pathways and a complex regulation of gene expression and enzymatic activities. We clearly need to know much more about these interactions before we can predict the effects of changing the expression of single genes on plant productivity.

## The Lipid Composition of Membranes Affects the Cell Biology and Physiology of Plants

The characterization of lipid mutants and the experimental modification of lipid composition by molecular-genetic means have now revealed some broad generalizations about the roles of membrane lipids in the cell biology and physiology of plants.

However, the biochemical and physical bases for the phenotypes that are observed remain uncertain. Changes in fatty acid composition have been shown to alter chloroplast size and architecture. Mutations that eliminate polyunsaturated fatty acids block photosynthesis and prevent the plant from growing autotrophically. Surprisingly, such mutant plants can grow on sucrose, indicating that photosynthesis is the only plant function that absolutely requires a polyunsaturated membrane. However, polyunsaturated fatty acids are also required as precursors of signalling compounds which are necessary for such diverse functions as pollen development and plant defence, such as jasmonic acid,. One of the most extensively studied issues in membrane biology is the relationship between lipid composition and the ability of organisms to adjust to temperature changes. For example, chill-sensitive plants experience sharp reductions in growth rate and development at temperatures between 0 and 12°C. The physical and physiological changes in chill-sensitive plants that are induced by exposure to low temperature, together with the subsequent expression of stress symptoms, are termed chilling injury. Many economically important crops, such as cotton, soybean, maize, rice, and many tropical and subtropical fruits, are classified as chill sensitive. In contrast, most plants of temperate origin are able to grow and develop at chilling temperatures and are classified as chill-resistant plants.

In attempts to link the biochemical and physiological changes associated with chilling injury with a single "trigger" or site of damage, it has been suggested that the primary event of chilling injury is a transition from a liquid-crystalline phase to a gel phase in the cellular membranes. According to this proposal, the transition from liquid-crystalline phase to gel phase would result in alterations in the metabolism of chilled cells and lead to injury and death of the chill-sensitive plants. The degree of unsaturation of the fatty acids would determine the temperature at which such damage occurs. A similar hypothesis has been proposed for chloroplast membranes in which the levels of disaturated phosphatidylglycerol (molecules containing no *cis* double bonds in the fatty acid chains) would determine the chilling sensitivity of plant species. However, the most recent research indicates that the relationship of membrane unsaturation to plant temperature responses is more subtle and complex than is suggested in these earlier hypotheses. On the one hand, studies of five different *Arabidopsis* mutants have demonstrated that reduced unsaturation can result in plants that grow well at 22°C but are less robust than wild-type plants when grown at 2 to 5°C.

These results were observed even though the lipid changes in most of the mutants are insufficient to cause a lipid phase transition. In addition, the responses of these mutants to low temperature appear quite distinct from classic chilling sensitivity, suggesting that normal chilling injury may not be related to the level of unsaturation of membrane lipids.

In one particular mutant, *fab1*, saturated forms of phosphatidylglycerol account for 43% of the total leaf phosphatidylglycerol—a higher percentage than is found in many chill-sensitive plants. Nevertheless, the mutant was completely unaffected (when compared with wild-type controls) by a range of low-temperature treatments that quickly led to the death of cucumber and other chill-sensitive plants. A complementary series of experiments has been carried out in tobacco, which is a chill-sensitive plant. The transgenic expression of exogenous genes in tobacco has been used specifically to decrease the level of saturated phosphatidylglycerol or to bring about a general increase in membrane unsaturation.

In each case, damage caused by chilling was alleviated to some extent. These new findings make it clear that the extent of membrane unsaturation or the presence of particular lipids, such as disaturated phosphatidylglycerol, can affect the responses of plants to low temperature. However, membrane lipid composition is not the major determinant of chilling sensitivity in plants.

## Utilization of Oil Reserves in Cotyledons

Although the pathway for the mobilization of triacylglycerols has been best characterized in castor bean, it seems to be similar in the storage tissues of other oilseeds.

However, not all seeds quantitatively convert fat to sugar. In castor bean, the endosperm degenerates after the fat and protein reserves are fully utilized. In many oilseeds—such as sunflower (*Helianthus annuus*), cotton (*Gossipium hirsutum*), and members of the squash family (Cucurbitaceae)—the cotyledons become green and photosynthesize after the food reserves are used up.

In these tissues only part of the stored lipid is converted to exported carbohydrate. Much of the lipid-derived carbon remains in the cotyledons, where it contributes to the synthesis of chloroplasts and other cellular structures. Acetyl-CoA is also used directly for the production of energy through respiration. As the greening process takes place, there is a transition in the peroxisome population of these cells: Some peroxisomes have fewer of the characteristics of glyoxysomes and more of leaf-type peroxisomes.

Such a transition is in keeping with the decreased requirement for the breakdown of stored lipids and the increased need to metabolize the products of photorespiration as the tissue goes from a heterotrophic to a more autotrophic mode of metabolism.

# Assimilation of Mineral Nutrients

### *Development of a Root Nodule*

The establishment of symbiosis between legumes and rhizobia involves the activation of genes in both the host and the symbiont and an elaborated exchange of signals. The formation of a root nodule, the specialized organ from a plant host that contains the symbiotic nitrogen fixing rhizobia involves two simultaneous processes: infection and nodule organogenesis. The different stages in the initiation and development of a root nodule are shown in Figure below. A detailed description of the infection process.

A recent study of the early infection stages and nodule development in two *Medicago* species and the symbiont, *Rhisobium meliloti* showed that extensive reorganization of microtubules occurred early in the infection process in the pericycle and the inner cortex where the nodule primordium forms. The changes in microtubule organization were observed by immunolocalization, and fluorescence and laser confocal microscopy. These findings show that upon infection, the pericycle is activated first, followed by the activation of the inner cortex, where the nodule primordium forms. These changes are followed by a major reorganization of the microtubular cytoskeleton network in outer tissues, associated with root hair activation and curling, the formation of pre-infection threads, and the initiation and the growth of an infection network.

### Measurement of Nitrogen Fixation

Besides $N_2$, the nitrogenase enzyme can reduce other substrates such as cyanide, azide and acetylene.

Direct measurement of $N_2$ fixation (reduction) requires a mass spectrometer, instrumentation that is not readily available, but the reduction of acetylene to ethylene can easily be measured by gas chromatography. Acetylene reduction involves two electrons, $C_2H_2 + 2e^- + 2\ H^+ \rightarrow C_2H_4$, as opposed to the eight electrons required for the reduction of $N_2$ and 2 $H^+$:

$$N_2 + 8e^- + 8H^+ + 16ATP \rightarrow 2NH_3 + H_2 + 16ADP + 16P_i$$

Therefore, reduction of four ethylene molecules correspond to the reduction of one $N_2$ molecule. The acetylene method has limitations, however: Acetylene blocks gas diffusion into nodules, and rhizosphere microorganisms produce ethylene. For these reasons, the acetylene method is suitable for comparative studies, while mass spectroscopy is used for direct measurements of $N_2$ reduction, which is required for precise quantification of nitrogen fixation.

## The Synthesis of Methionine

Methionine and cysteine are the two sulfur-containing amino acids found in proteins. Methionine is synthesized in plastids from cysteine. In the first step of the pathway, cysteine and O-phosphohomoserine react to form cystathionine via the enzyme cystathionine-γ-synthase. Cystathionine is cleaved into homocysteine, pyruvate, and ammonia by cystathionine-β-lysase. Finally, methionine synthase methylates homocysteine to form methionine (Lea 1997).

# Chapter 3

# Photosynthesis

## Principles of Spectrophotometry

Much of what we know about the photosynthetic apparatus was learned through spectroscopy—that is, measurements of the interaction of light and molecules. Spectrophotometry is an important branch of spectroscopy that focuses on the technique of measurement. Here we will examine four topics: Beer's law, the measurement of absorbance, action spectra, and difference spectra.

### *Beer's Law*

An essential piece of information about any molecular species is how much of it is present. Quantitative measures of concentration are one of the cornerstones of biological science. Of all the methods that have been devised for measuring concentration, by far the most widely applied is absorption spectrophotometry.

In this technique, the amount of light that a sample absorbs at a particular wavelength is measured and used to determine the concentration of the sample by comparison with appropriate standards or reference data. The most useful measure of light absorption is the absorbance ($A$), also commonly called the optical density (OD). The absorbance is defined as $A = \log I_0 / I$ where $I_0$ is the intensity of light that is incident on the sample and $I$ is the intensity of light that is transmitted by the sample. The absorbance of a sample can be related to the concentration of the absorbing species through Beer's law:

$$A = \varepsilon cl$$

where $c$ is concentration, usually measured in moles per litter; $l$ is the length of the light path, usually 1 cm; and $\varepsilon$ is a proportionality constant known as the molar extinction coefficient, with the units of litres per mole per centimetre. The value of $\varepsilon$ is a function of both the particular compound being measured and the wavelength. Chlorophylls typically have an $\varepsilon$ value of about 100,000 L $mol^{-1}$ $cm^{-1}$. When more than one component of a complex mixture absorbs at a given wavelength, the absorbances due to the individual components are generally additive.

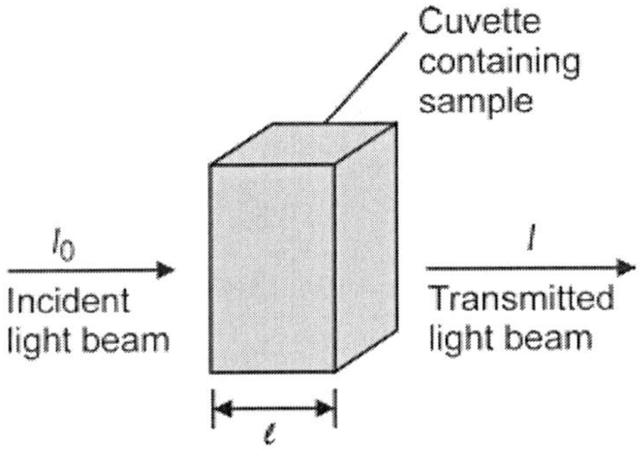


***Figure:*** *Definition of absorbance. A monochromatic incident light beam of intensity $I_0$ traverses a sample contained in a cuvette of length (l). Some of the light is absorbed by the chromophores in the sample, and the intensity of light that emerges is I.*

## The Spectrophotometer

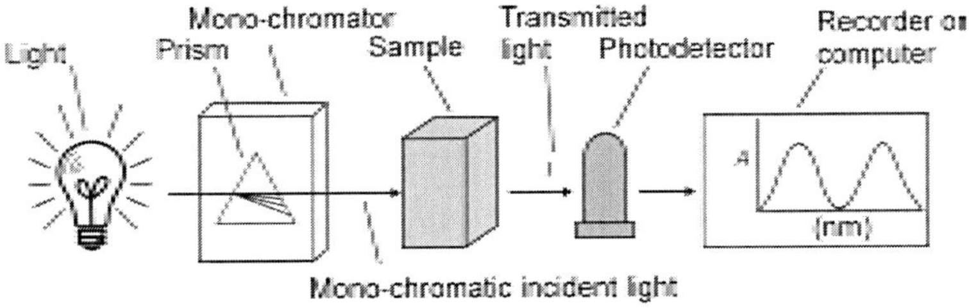


***Figure:*** *Schematic diagram of a spectrophotometer. The instrument consists of a light source, a monochromator that contains a wavelength selection device such as a prism, a sample holder, a photodetector, and a recorder or computer. The output wavelength of the monochromator can be changed by rotation of the prism; the graph of absorbance versus wavelength is called a spectrum.*

The absorbance is measured by an instrument called a spectrophotometer. The essential parts of a spectrophotometer include a light source, a wavelength selection device such as a monochromator or filter, a sample chamber, a light detector, and a readout device, usually also include a computer, which is used for storage and analysis of the spectra. The most useful machines scan the wavelength of the light that is incident on the sample and produce, as output, spectra of absorbance versus wavelength.

### *Action Spectra*

The use of action spectra has been central to the development of our current understanding of photosynthesis. An action spectrum is a graph of the magnitude of the biological effect observed as a function of wavelength. Examples of effects measured by action spectra are oxygen evolution and hormonal growth responses due to the action of phytochrome. Often an action spectrum can identify the chromophore responsible for a particular light-induced phenomenon. Action spectra were instrumental in the discovery of the existence of the two photosystems in $O_2$-evolving photosynthetic organisms.

Some of the first action spectra were measured by T. W. Engelmann in the late 1800s. Engelmann used a prism to disperse sunlight into a rainbow that was allowed to fall on an aquatic algal filament. A population of $O_2$-seeking bacteria was introduced into the system.

The bacteria congregated in the regions of the filaments that evolved the most $O_2$. These were the regions illuminated by blue light and red light, which are strongly absorbed by chlorophyll. Today, action spectra can be measured in room-sized spectrographs in which the scientist enters a huge monochromator and places samples for irradiation in a large area of the room bathed by monochromatic light. But the principle of the experiment is the same as that of Engelmann's experiments.

Schematic diagram of the action spectrum measurements by T. W. Engelmann. Engelmann projected a spectrum of light onto the spiral chloroplast of the filamentous green alga *Spirogyra* and observed that oxygen-seeking bacteria introduced into the system collected in the region of the spectrum where chlorophyll pigments absorb. This action spectrum gave the first indication of the effectiveness of light absorbed by accessory pigments in driving photosynthesis.

### *Difference Spectra*

An important technique in studies of photosynthesis is light-induced difference spectroscopy, which measures changes in absorbance. In this technique, bright light, often called actinic light, is used to illuminate a sample, while a dim beam of light is used to measure the absorbance of the sample at wavelengths other than that of the actinic beam. In this way a difference spectrum is obtained, which represents the changes in the absorption spectrum of the sample induced by illumination with the actinic light.

Absorption bands that disappear upon illumination appear as negative peaks; new bands that appear upon illumination appear as positive peaks. Difference spectra give important clues to the identity of molecular species participating in the photoreactions of photosynthesis.

By the use of special flash techniques, it is possible to record the difference spectrum at a given time after flash excitation. Multiple difference spectra recorded at different times after flash excitation can be used to measure the kinetics of the chemical reactions that follow photon excitation of a reaction centre. These techniques can have extraordinary time resolution, in some cases less than a picosecond ($10^{-12}$ s), and have provided great insights into the earliest events in the photosynthetic energy storage process.

## Antagonistic Effects of Light on Cytochrome Oxidation

The puzzling red drop and enhancement effects were explained by experiments performed by Louis Duysens of the Netherlands. He suggested that two phtotchemical events were responsible for enhancement effects. One photochemical event produced an oxidation while the other produced a reduction. Chloroplasts contain cytochromes, iron-containing proteins that function as intermediate electron carriers in photosynthesis. Duysens found that when a sample of a red alga was illuminated with long-wavelength light, the cytochrome became mostly oxidized. If light of a shorter wavelength was also present, the effect was partly reversed. These antagonistic effects can be explained by a mechanism involving two photochemical events: one that tended to oxidize the cytochrome and one that tended to reduce it.

We know now that in the red region of the spectrum, one of the photoreactions, known as photosystem I (PS-I), absorbs preferentially far-red light of wavelengths greater than 680 nm, while the second, known as photosystem II (PS-II), absorbs red light of 680 nm well and is driven very poorly by far-red light. This wavelength dependence explains the enhancement effect and the red drop effect. Another difference between the photosystems is that Photosystem I produces a strong reductant, capable of reducing $NADP^+$, and a weak oxidant. Photosystem II produces a very strong oxidant, capable of oxidizing water, and a weaker reductant than the one produced by photosystem I. This reductant of photosystem II rereduces the oxidant produced by photosystem I, which explains the antagonistic effect.

### Structures of Two Bacterial Reaction Centres

In 1984, Hartmut Michel, Johann Deisenhofer, Robert Huber, and coworkers in Munich solved the three-dimensional structure of the reaction centre from the purple photosynthetic bacterium *Rhodopseudomonas viridis*. This landmark achievement, for which a Nobel prize was awarded in 1988, was the first high-resolution X-ray structure determination for an integral membrane protein and the first structure determination for a reaction centre complex.

The protein part of the complex consists of four separate polypeptides. Two of them, called L and M (for *l*ight and *m*edium mass) bind all of the bacteriochlorophyll, quinone, and carotenoid cofactors of the complex.

The structure has a twofold symmetry about an axis perpendicular to the plane of the membrane, hinting at a dimeric nature of the reaction centre. The ten transmembrane portions of the L and M peptides (five from each) are arranged in α helices, and there are almost no charged amino acid residues in the interior of the membrane. The H (*h*eavy) protein has a single transmembrane helix and is localized mostly on the cytoplasmic side of the membrane. The C (cytochrome) subunit is located in the periplasmic region (the region between the bacterial plasma membrane and the outer membrane). The geometric arrangement of the pigments and the quinones (electron acceptors), with the protein removed. A similar arrangement is found in the reaction centre of another purple photosynthetic bacterium, *Rhodobacter sphaeroides*, except that the C subunit is not present. Two of the bacteriochlorophyll molecules are in intimate contact with each other and are known as the special pair. This dimer, whose existence was predicted from magnetic-resonance studies, is the photoactive portion of the complex. An electron is transferred from this dimer along the sequence of electron carriers on the right side of the complex. Detailed analysis of these structures, along with analysis of numerous mutants, has revealed many of the principles involved in the energy storage processes that are carried out by all reaction centres.

The bacterial reaction centre structure is thought to be similar in many ways to that found in photosystem II from oxygen-evolving organisms, especially in the electron acceptor portion of the chain. The proteins that make up the core of the bacterial reaction centre are relatively similar in sequence to their photosystem II counterparts, implying an evolutionary relatedness.

## Midpoint Potentials and Redox Reactions

Redox reactions, midpoint potentials, and their relationship to the laws of thermodynamics. These concepts are useful for our discussion of electron flow from $H_2O$ to $NADP^+$ and the interactions between the different electron carriers.

The midpoint potential ($E_m$) is a measure of the tendency of a compound to take electrons from other compounds. A large positive midpoint potential means that the compound is a strong oxidant; a large negative value means that the compound is a strong reductant (in both cases relative to the standard hydrogen electrode). Equilibrium constants can easily be predicted from midpoint potentials, in the same way that free energies were related to equilibrium constants. Midpoint potentials for many chemical and biochemical reactions have been measured and tabulated. The y-axis on the Z scheme, midpoint potentials of the electron carriers, with negative values higher than positive ones. This choice makes reactions that are spontaneous (releasing free energy) appear "downhill" on the graph.

Knowledge of the midpoint potentials of the various electron carriers is important in establishing the pathway of electron flow in any biochemical electron transport system, such as those found in chloroplasts or mitochondria. Researchers make this measurement usually by carrying out a redox titration. They adjust, or poise, the sample at a particular redox potential, usually by adding small amounts of oxidants or reductants. Redox mediators, small molecules that permit rapid equilibration between the sample and the electrodes of the measurement system, must be included to ensure that the system is at equilibrium when the measurement is made. Several measurements are made at a variety of redox potentials. The sample is stirred in a special cell that contains platinum and reference electrodes, and chemical oxidants and reductants are added to adjust the redox potential, which is read with a voltmeter.

We can measure the extent of the redox reaction by following a particular property of the sample, usually absorbance, at each potential. In the example illustrated, the reduced form of the compound has an absorbance that decreases as the compound is oxidized. The fraction of reduced form at each potential is plotted against redox potential, and the midpoint potential ($E_m$) is determined as the potential at which the compound is half oxidized and half reduced.

## Oxygen Evolution

The chemical mechanism of photosynthetic water oxidation is not yet known, although there is a great deal of indirect evidence about the process. If a sample of dark-adapted photosynthetic membrane is exposed to a sequence of very brief, intense flashes, a characteristic pattern of oxygen production is observed.

Little or no oxygen is produced on the first two flashes, and maximal oxygen is released on the third flash and every fourth flash thereafter, until eventually the yield per flash damps to a constant value. This remarkable result was first observed by Pierre Joliot in the 1960s.

A schematic model explaining these observations, proposed by Kok and coworkers, has been widely accepted. This model for the photooxidation of water, called the S state mechanism, consists of a series of five states, known as $S_0$ to $S_4$, which represent successively more oxidized forms of the water-oxidizing enzyme system, or oxygen-evolving complex. The light flashes advance the system from one S state to the next, until state $S_4$ is reached. State $S_4$ produces $O_2$ without further light input and returns the system to $S_0$. Occasionally, a centre does not advance to the next S state upon flash excitation, and less frequently, a centre is activated twice by a single flash. These misses and double hits cause the synchrony achieved by dark adaptation to be lost and the oxygen yield eventually to damp to a constant value. After this steady state has been reached, a complex has the same probability of being in any of the states $S_0$ to $S_3$ ($S_4$ is unstable and occurs only transiently), and the yield of $O_2$ becomes constant. States $S_2$ and $S_3$ decay in the dark, but only as far back as $S_1$, which is stable in the dark. Therefore, after adaptation to the dark, approximately three-fourths of the oxygen-evolving complexes appear to be in state $S_1$ and one-fourth in state $S_0$. This distribution of states explains why the maximum yield of $O_2$ is observed after the third of a series of flashes given to dark-adapted chloroplasts.

## Photosystem I

The PS-I reaction centre is composed of a multiprotein complex. The reaction centre chlorophyll P700 and about 100 core antenna chlorophylls are bound to two proteins, PsaA and PsaB, with molecular masses in the range of 66 to 70 kDa.

PS-I reaction centre complexes have been isolated from several organisms and found to contain the 66 to 70 kDa proteins, along with a variable number of smaller proteins in the range of 4 to 25 kDa. Some of these proteins serve as binding sites for the soluble electron carriers plastocyanin and ferredoxin.

The functions of some of the other proteins are not well understood. An 8-kDa protein contains some of the bound iron-sulfur centres that serve as early electron acceptors in photosystem I. The structure of the PS-I complex from pea has been determined to a resolution of 4.4 E, and the positions of many of the chlorophylls and electron transfer components have been located. In their reduced form, the electron carriers that function in the acceptor region of photosystem I are all extremely strong reducing agents. These reduced species are very unstable and thus difficult to identify. Evidence indicates that one of these early acceptors is a chlorophyll molecule, and another is a quinone species, phylloquinone, also known as vitamin $K_1$.

Additional electron acceptors include a series of three membrane-associated iron–sulfur proteins, or bound ferredoxins, also known as Fe–S centres Fe–$S_X$, Fe–$S_A$, and Fe–$S_B$. Fe–S centre Fe–$S_X$ is part of the P700-binding protein; centres Fe–$S_A$ and Fe–$S_B$ reside on an 8 kDa protein that is part of the PS-I reaction centre complex. Electrons are transferred through centres Fe–$S_A$ and Fe–$S_B$ to ferredoxin, a small, water-soluble iron–sulfur protein. The membrane-associated flavoprotein ferredoxin–NADP reductase (FNR) reduces $NADP^+$ to NADPH, thus completing the sequence of noncyclic electron transport that begins with the oxidation of water. Electron transport between PSII and PSI is mediated by plastohydroquinone, the cytochrome $b_6f$ complex, and plastocyanin. The cytochrome $b_6f$ complex contains two *b*-type hemes and one *c*-type heme.

## ATP Synthase

The chemical mechanism by which ATP is synthesized is not yet understood in detail. However, considerable evidence now supports a mechanism, first proposed by Paul Boyer, in which the principal energy-requiring step is the release of bound ATP from the enzyme.

It has also been proposed that during catalysis a large portion of the $CF_1$ complex rotates about a bearing consisting of the $\gamma$ subunit. The $\gamma$ subunit may act as a camshaft does, rotating alternately against the $\alpha$ and $\alpha$ subunits. The energy of the conformational movements is then translated into phosphoanhydride bond energy. The thylakoid ATP synthase has two segments: $CF_0$ is a transmembrane segment, and $CF_1$ is a hydrophilic segment at the stomal surface. $CF_0$ traslocates protons from the chloroplast lumen to the catalytic part of the enzyme and $CF_1$ cataylses the conversion of ADP and inorganic phosphate to ATP.

The chloroplast ATP synthase has nine different subunits. $CF_1$ is made of two large subunits, α and β, plus three smaller subunits, γ, δ, and ε. Each $CF_1$ molecule has three copies of the α and β subunits and one copy of the γ, δ and ε subunits. The α and β subunits bind ADP and phosphate and catalyize the phosphorylation of ADP into ATP.

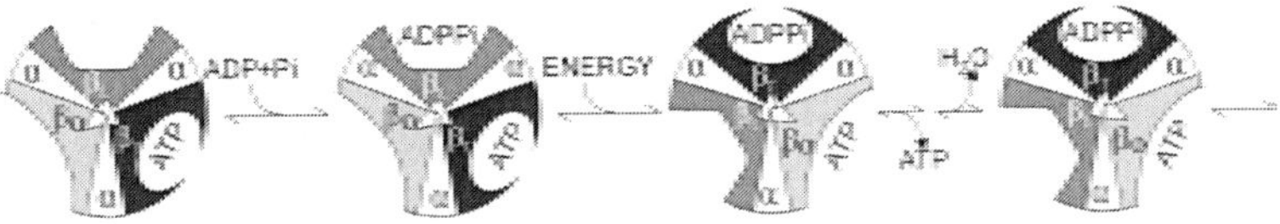


***Figure:*** *The model of Boyer's binding change mechanism. The a and b subunits configure three nucleotide binding sites: O, which provides the early binding site for ADP and inorganic phosphate, L, to which the ADP and inorganic phosphate bind after migrating from O, and T, which tightly binds ATP. Energy ensuing from the movement of protons from the chloroplast lumen to the stroma drives the rotation of the g subunit of $CF_1$, and the interconversion of the binding sites and the release of an ATP molecule.*

Studies on the mitochondrial ATP synthase have increased our understanding of how the ATP synthases work. As proposed by Paul Boyer in 1997, ATP synthesis takes place by a binding change mechanism. According to this model, the energy of the proton gradient is used to release a tightly bound form of ATP from a catalytic binding side of the enzyme. As shown in Figure below, $CF_1$ has three nucleotide binding sites, L, T and O. ADP and inorganic phosphate are postulated to first bind to the O site, and then move to the L site. The T site binds ATP. As protons move from the lumen to the stromal region through the $CF_0$ channel, energy is released and the γ subunit of $CF_1$ rotates. The rotation causes conformation changes in the three nucleotide binding sites, which interconvert, thus changing the afiity of the sites for the nucleotides. As the T site converts into a O site, ATP is released and another cycle starts. Direct observation of the rotation of the γ subunit of $CF_1$ has provided strong experimental support for the binding change mechanism.

## Mode of Action of Some Herbicides

Herbicides of one major class—about half of the commercially important compounds—act by interrupting photosynthetic electron flow.

A shows the chemical structure of two of these compounds. The precise sites of action of many of these agents have been found to lie either at the reducing side of photosystem I (for example, in paraquat) or in the quinone acceptor complex in the electron transport chain between the two photosystems.

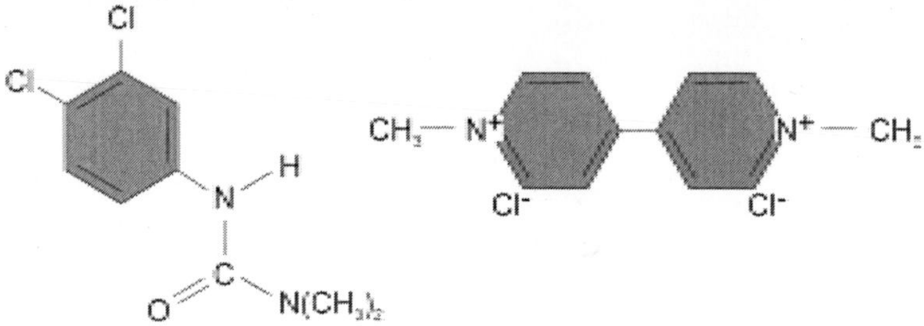


***Figure:*** *The use of herbicides to kill unwanted plants is widespread in modern agriculture. Many different classes of herbicides have been developed, and they act by blocking amino acid, carotenoid, or lipid biosynthesis or by disrupting cell division. Understanding the mode of action of herbicides has been an important tool in research on plant metabolism and has facilitated their application under different agricultural practices (Ashton and Crafts 1981). Here, the chemical structure of two herbicides that block photosynthetic electron flow are shown. DCMU is also known as diuron. Paraquat has acquired public notoriety because of its use on marijuana crops.*

Paraquat acts by intercepting electrons between the bound ferredoxin acceptors and NADP and then reducing oxygen to superoxide ($O_2^-$). Superoxide is a free radical that reacts nonspecifically with a wide range of molecules in the chloroplast, leading to the rapid loss of chloroplast activity. Lipid molecules in cell membranes are especially sensitive. The herbicides that act on the quinone acceptor complex compete with plastoquinone for the $Q_B$ binding site. If herbicide is present, it displaces the oxidized form of plastoquinone and occupies the specific binding site for the quinone acceptor, which is thought to lie on the D1 herbicide-binding protein. The herbicide is not able to accept electrons, so the electron is unable to leave $Q_A$, the first quinone acceptor. Thus, the binding of herbicide effectively blocks electron flow and inhibits photosynthesis. Many herbicides that act in this manner also inhibit electron flow in photosynthetic bacteria that have quinone-type electron acceptor complexes.

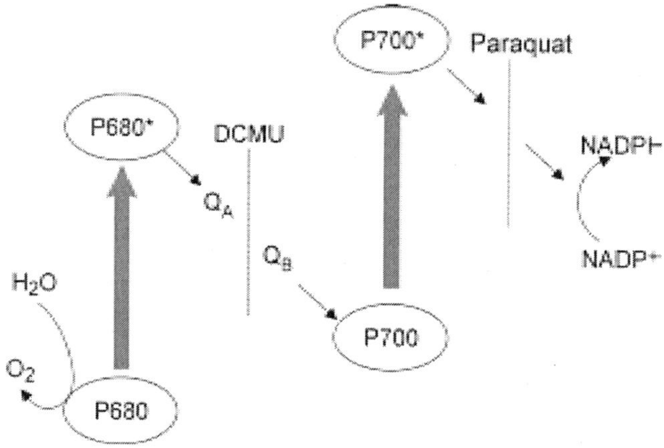


***Figure:*** *Sites of action of herbicides. Many herbicides, such as DCMU, act by blocking electron flow at the quinone acceptors of photosystem II, by competing for the binding site of plastoquinone that is normally occupied by $Q_B$. Other herbicides, such as paraquat, act by accepting electrons from the early acceptors of photosystem I and then reacting with oxygen to form superoxide, $O_2^-$, a species that is very damaging to chloroplast components, especially lipids.*

In recent years, herbicide-resistant biotypes of common weeds have appeared in areas where a single type of herbicide has been used continuously for several years. These biotypes can be orders of magnitude more resistant to certain classes of herbicides than the nonresistant plant is. In several cases the resistance factor has been traced to a single amino acid substitution in the D1 protein. This change, presumably in the quinone (and herbicide) binding region of the peptide, lowers the binding affinity of the herbicide, making it much less effective.

The possibility of fine-tuning the herbicide sensitivity of crop plants by making subtle changes in the proteins of the PS-II reaction centre has created a great deal of interest in the agricultural chemical industry. Through biotechnology, it is now possible to make a crop plant that is resistant to a particular herbicide, which can then be applied to control undesirable plants that are not resistant. The success of this approach will depend on whether undesirable side effects of the herbicide-resistant mutations can be controlled and on how rapidly the weeds acquire resistance through natural selection or gene transfer. Thus, it is a race between science and evolutionary change.

A complementary approach is to insert resistance factors to certain pests, such as insects, into the plant via genetic engineering. The plants then produce the toxic species themselves, eliminating the need for insecticides.

### *Chlorophyll Biosynthesis*

In the first phase of chlorophyll biosynthesis, the amino acid glutamic acid is converted to 5-aminolevulinic acid (ALA). This reaction is unusual in that it involves a covalent intermediate in which the glutamic acid is attached to a transfer RNA molecule. This is one of a very small number of examples in biochemistry in which a tRNA is utilized in a process other than protein synthesis. Two molecules of ALA are then condensed to form porphobilinogen (PBG), which ultimately form the pyrrole rings in chlorophyll. The next phase is the assembly of a porphyrin structure from four molecules of PBG. This phase consists of six distinct enzymatic steps, ending with the product protoporphyrin IX.

All the biosynthesis steps up to this point are the same for the synthesis of both chlorophyll and heme. But here the pathway branches, and the fate of the molecule depends on which metal is inserted into the centre of the porphyrin. If magnesium is inserted by an enzyme called magnesium chelatase, then the additional steps needed to convert the molecule into chlorophyll take place; if iron is inserted, the species ultimately becomes heme.

The next phase of the chlorophyll biosynthetic pathway is the formation of the fifth ring (ring E) by cyclization of one of the propionic acid side chains to form protochlorophyllide. The pathway involves the reduction of one of the double bonds in ring D, using NADPH. This process is driven by light in angiosperms and is carried out by an enzyme called protochlorophyllide oxidoreductase (POR). Non-oxygen-evolving photosynthetic bacteria carry out this reaction without light, using a completely different set of enzymes. Cyanobacteria, algae, lower plants, and gymnosperms contain both the light-dependent POR pathway and the light-independent pathway.

Seedlings of angiosperms grown in complete darkness lack chlorophyll, because the POR enzyme requires light. These *etiolated* plants very rapidly turn green when exposed to light. The final step in the chlorophyll biosynthetic pathway is the attachment of the phytol tail, which is catalyzed by an enzyme called chlorophyll synthetase.

The elucidation of the biosynthetic pathways of chlorophylls and related pigments is a difficult task, in part because many of the enzymes are present in low abundance. Recently, genetic analysis has been used to clarify many aspects of these processes.

## Photosynthesis: Carbon Reactions

### *How the Calvin Cycle Was Elucidated*

The elucidation of the Calvin cycle was the result of a series of elegant experiments by Melvin Calvin and his colleagues in the 1950s, for which a Nobel prize was awarded in 1961. They used suspensions of the unicellular eukaryotic green alga *Chlorella* to trace the path of carbon. The Calvin cycle is also referred to as the reductive pentose phosphate cycle to distinguish the photosynthetic cycle from the oxidative pentose phosphate pathway, with which it shares several enzymes. Yet another designation of the Calvin cycle is photosynthetic carbon reduction cycle.

To elucidate the cycle, the investigators first exposed the algal cells to constant conditions of light and $CO_2$ to establish steady-state photosynthesis.

Next they added radioactive $^{14}CO_2$ for a brief period to label the intermediates of the cycle. They then killed the cells and inactivated their enzymes by plunging the suspension into boiling alcohol. They separated the $^{14}C$-labeled compounds from one another and identified them by their positions on two-dimensional paper chromatograms.

In this manner, 3-phosphoglycerate and the various sugar phosphates were identified as intermediates in carbon fixation. By exposing the cells to $^{14}CO_2$ for progressively shorter periods of time, Calvin's group was able to identify 3-phosphoglycerate as the first stable intermediate.

Therefore, the other labelled sugar phosphates must be derived from a subsequent reduction of 3-phosphoglycerate.

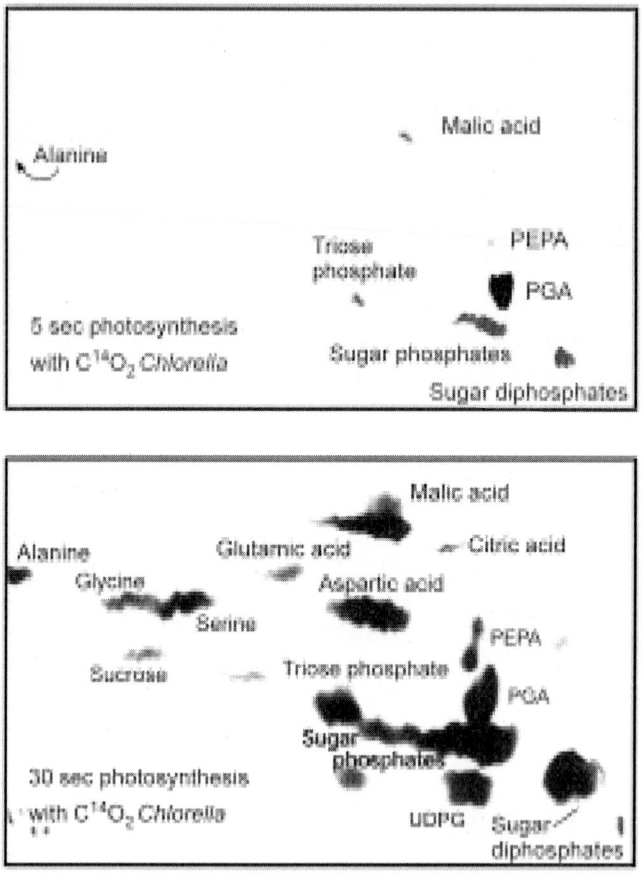


***Figure:** Autoradiograms showing the labelling of carbon compounds in the alga Chlorella after exposure to $^{14}CO_2$. The time intervals shown in the figures indicate the length of exposure to the radiolabel. At the indicated time intervals, the reaction was terminated, by plunging of the contents into boiling alcohol. The labelled compounds in the cell homogenates were then separated by paper chromatography. The heavy labelling of 3-phosphoglycerate (PGA) after the shorter exposure indicates that it is the first stable intermediate of the Calvin (reductive pentose phosphate) cycle.*

To deduce the path of carbon, it was necessary to determine the distribution of $^{14}C$ in each of the labelled sugars. After a brief exposure to $^{14}CO_2$, individual intermediates were isolated and chemically degraded so that the amount of $^{14}C$ in each carbon atom could be determined. The results showed that 3-phosphoglycerate was initially labelled predominantly in the carboxyl group. This finding suggested that the initial $CO_2$ acceptor was a two-carbon compound, and it prompted a long and futile search for such a compound.

The subsequent discovery that pentose monophosphates and a pentose bisphosphate (ribulose) participate in the cycle raised the possibility that the initial $CO_2$ acceptor was a five-carbon compound. This conceptual breakthrough rapidly led to the identification of ribulose-1,5-bisphosphate as the $CO_2$ acceptor and to the formulation of the complete cycle. The operation of the cycle explained the $^{14}C$-labeling pattern of the other intermediates. To demonstrate conclusively that a metabolic pathway exists, it is necessary to prove that the postulated enzymes catalyze the proposed reactions in the test tube. Ideally, the rates of these enzyme reactions (in vitro) should be equal to or in excess of those observed in the intact cell (in vivo). However, this evidence can be used only to support a proposed pathway. Failure to demonstrate a reaction in vitro does not prove that the reaction does not occur.

All of the reactions of the Calvin cycle have been demonstrated in vitro. With one exception, the in vitro rates of enzymatic activity are well in excess of the maximal observed rates of photosynthesis. That exception is rubisco, which, when assayed under the levels of $CO_2$ and $O_2$ in air, shows barely enough in vitro activity to account for the observed rate of photosynthesis in air. Although great progress has been made in increasing activity by providing proper conditions for activation, researchers continue to debate whether rubisco is a rate-limiting step in photosynthesis.

## Rubisco: A Model Enzyme for Studying Structure and Function

With the exception of the carbon fixed by a some prokaryotic organisms, most of the carbon fixed on Earth is processed by the Calvin cycle. The concentration of rubisco, the carboxylating enzyme of the Calvin cycle, is generally high. For example, it accounts for 50% or more of the total protein in plant leaves, and its concentration within chloroplasts is extremely high (*ca.* 0.2 g/ml).

Rubisco occurs in two functionally analogous forms, having different structure, distribution, and $O_2$ sensitivity. Oxygenic phototroph (cyanobacteria, chloroplasts from high plants) and many photosynthetic bacteria contain a form of the enzyme, (form I), made up of eight large (L) catalytic subunits (about 55 kDa each) and eight small (S) subunits (about 14 kDa each), giving a molecular mass of about 560 kDa for the complete protein ($L_8S_8$). The 8 L subunits of form I are arranged as an octameric core surrounded by two layers of four S subunits, with each layer located on opposite sides of the molecule.

Form II of rubisco is found in some photosynthetic purple nonsulfur bacteria, and it is composed of two L subunits, each 50 kDa. Eukaryotic algal dinoflagellates also use a dimeric enzyme of proteobacterial ancestry. A rubisco-like enzyme lacking catalytic activity has recently been identified in photosynthetic green sulfur bacteria which lack the Calvin cycle. The evidence suggests that the enzyme may function in sulfur oxidation to provide electrons for the photosynthetic electron transport chain. Genome sequences show that the rubisco-like protein also occurs in certain heterotrophic bacteria.

### *Rubisco Activase*

Rubisco catalyzes the incorporation of $CO_2$ to the C-2 of the enediol form of ribulose-1,5-bisphosphate and the subsequent cleavage of the unstable intermediate to yield two molecules of 3-phosphoglycerate The conversion of the inactive form of rubisco to an active state, which is required for both the carboxylation and oxygenation reactions, requires the carbamylation of a specific lysine residue (in spinach, Lys-201). As a consequence of the binding of the $CO_2^-$ molecule to the lysine residue, the enzyme acquires a triad of anionic residues that provide the binding site for the cofactor $Mg^{2+}$ (in spinach, the other two residues are Asp-203 and Glu-204).

At this stage, the incorporation of ribulose-1,5-bisphosphate causes conformational changes in rubisco that allows the enzyme to sequester the sugar bisphosphate from the solvent and, upon the binding of a $CO_2$ or $O_2$ molecule, to start the catalytic cycle. However, the tight binding of ribulose-1,5-bisphosphate to the uncarbamylated enzyme molecules displaces the equilibrium to the dead end complex, rubisco-ribulose-1,5-bisphosphate, so that the rate of reaction declines rapidly until it finally stops. How is this problem solved?

A combination of genetic, physiological and biochemical studies have provided the solution to the long-term riddle of rubisco activation. In vivo, some mutants of *Arabidopsis thaliana* lack light-mediated activation of rubisco and can only grow at high concentrations of $CO_2$. Surprisingly, the mutation does not impair the response of the enzyme to modulation by $CO_2$ and $Mg^{2+}$ in vitro. These contradictory results led to the discovery of a protein, rubisco-activase, that uncouples ribulose-1,5-bisphosphate from decarbamylated active sites and, in so doing, promotes the access of $CO_2$ and $Mg^{2+}$ for the carbamylation of the enzyme. The importance of this process is that the rate at which rubisco-activase enhances the catalytic capacity of rubisco is linked to the hydrolysis of ATP (50 molecules of ATP hydrolyzed / 1 rubisco site activated).

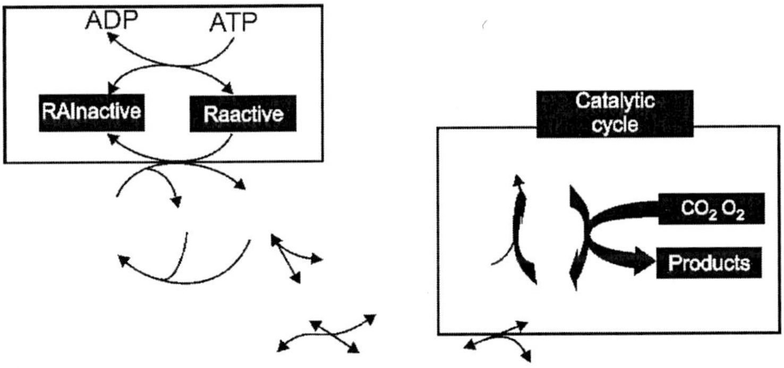


***Figure:*** *Modulation of rubisco by rubisco-activase. The combination or the interaction of rubisco (E) with modulators distributes the enzyme among different states (yellow boxes). The unprotonated form of a specific lysine residue (E-$NH_2$) reacts with $CO_2$ releasing a proton and forming the carbamate (E-NH-$CO_2^-$) that is stabilized by the cofactor $Mg^{2+}$ (E-NH-$CO_2^-$. $Mg^{2+}$). At this stage (blue box), rubisco is catalytically competent and thereby binds ribulose-1,5-bisphosphate (RuBP) yielding the ternary complex (E-NH-$CO_2^-$. $Mg^{2+}$. RuBP) in which the latter is converted to the enediol form. This entails the incorporation of $CO_2$ or $O_2$ in the active site to generate the products of carboxylation or oxygenation, respectively. The high affinity of ribulose-1,5-bisphosphate for free rubisco (E-$NH^{3+}$) generates the dead-end complex (E-$NH^{3+}$. RuBP) thereby displacing the above equilibria.*

Two isoforms of rubisco activase, (42 and 46 kDa) found in several species differ in the C-terminal region as result of the differential splicing of a common pre-mRNA. Both isoforms catalyze the activation of rubisco and the hydrolysis of ATP but the smaller isoform has higher capacity than the larger isoform for both activities. Evidence for the fact that the two processes are independent are provided by studies showing that the digestion of native rubisco-activase with trypsin or the modification of its primary structure by site-directed mutagenesis do not elicit the same responses in the stimulation of rubisco and the hydrolysis of ATP. More importantly, the deletion of some residues at the C-region of rubisco-activase doubles the rate of rubisco activation with only a minor effect on the hydrolysis of ATP. Rubisco activation by the activase requires a mutual recognition of both proteins. Rubisco-activase from tobacco (*Solanaceae*) fails to activate rubisco from spinach or *Chlamydomonas* (non-*Solanaceae*) and, similarly, the rubisco-activase from spinach or *Chlamydomonas* fails to stimulate rubisco from tobacco. However, the substitution of some residues on the surface of the large subunit of *Chlamydomonas* rubisco shifts the specificity from non-*Solanaceae* to *Solanaceae* rubisco-activase. These experiments underscore the contribution of specific amino acids to protein-protein interaction and provide experimental evidence for a recognition step separated from the subsequent activation process.

## Three Variations of $C_4$ Metabolism

### *The Carbon Path*

The primary carboxylation reaction, catalyzed by PEP carboxylase, which is common to all three variants, occurs in the cytosol of the mesophyll cells and uses $HCO_3^-$ rather than $CO_2$ as a substrate. In the NADP-ME variant, oxaloacetate is rapidly reduced to malate in the mesophyll chloroplasts by NADPH using NADP:malate dehydrogenase. The malate formed enters the chloroplast of the bundle sheath cell and there undergoes oxidative decarboxylation, yielding pyruvate. The $CO_2$ released within the bundle sheath cells is converted to carbohydrate by the Calvin cycle. The residual $C_3$ acid is transported back to the mesophyll as pyruvate and converted to phosphoenolpyruvate in the mesophyll chloroplast. This reaction takes place in the mesophyll chloroplast of all three $C_4$ variants. The cycle then starts again with carboxylation of the phosphoenolpyruvate to produce oxaloacetate. Another process localized in the mesophyll chloroplast of all three $C_4$ variants is the reduction of 3-phosphoglycerate to triose phosphates, which occurs within the Calvin cycle proper in $C_3$ chloroplasts. In contrast, all $C_4$ variants export 3-phosphoglycerate from the bundle sheath for reduction in the mesophyll chloroplast. After reduction, the triose phosphates are transported back to the bundle sheath chloroplasts, for use in the operation of the Calvin cycle there.

One possible reason for this remarkable split of the Calvin cycle between the bundle sheath and the mesophyll chloroplast is to restrict electron transport activity and oxygen evolution in the bundle sheath, and thus avoid oxygen levels that would favour photorespiration. The transport of metabolites of the $C_4$ pathway is accompanied by movements of protons and other ionic species in order to maintain pH and charge balance.

## Photorespiration in CAM Plants

Many terrestrial plants, adapted to hot and arid regions where water is scarce, are stressed by the severe effects of full sunlight, particularly drought and high temperature. To avoid high rates of water transpiration, the stomata of these species are tightly closed during the hot diurnal hours and perform most of their gas exchange with the surrounding atmosphere at night, when the air is relatively cool and humid. On the other hand, the vital process of photosynthesis in which $CO_2$ is incorporated into carbohydrates takes place during the day. To cope with the lack of an efficient diurnal influx of atmospheric $CO_2$, these species have evolved a different photosynthetic mechanism, called "crassulacean acid metabolism" (CAM).

This unusual process takes place within single leaf cells in which the uptake and subsequent assimilation of atmospheric $CO_2$ involves three compartments: cytosol, vacuole, and chloroplast. In general, CAM plants are common among desert succulents (e.g., *Cactaceae*) exposed to daytime high temperatures and low relative humidity, or among tropical forest epiphytes (*Bromeliaceae*) which are exposed to physiological drought under humid conditions because of their aerial growth habit. Remarkably, CAM also occurs in aquatic species (*Isoetaceae*) found worldwide in marshy and shallow habitats that exhibit a diurnal limitation of $CO_2$.

During the cool night-time hours, the open stomata facilitate the uptake of atmospheric $CO_2$ which diffuses into the cell cytosol, generating $HCO_3^-$. At this stage, phosphoenolpyruvate-carboxylase (PEPCase) catalyzes the incorporation of $HCO_3^-$ into phosphoenolpyruvate, yielding oxaloacetic acid and related four-carbon organic acids (particularly malic acid), that accumulate in the vacuole.

During the hot hours of the day, the stomatal closure characteristic of terrestrial CAM plants reduces water loss but causes an internal $CO_2$ shortage. To overcome this deficiency, the vacuole releases malic acid into the cytosol where the action of decarboxylating enzymes, NADP–malic enzyme or PEP-carboxykinase, produces pyruvate and $CO_2$. Inside the stroma of chloroplasts, Rubisco incorporates the released $CO_2$ into 3-phosphoglycerate which, in concert with the ATP and NADPH synthesized in the light via photophosphorylation, initiates a series of Benson–Calvin cycle reactions that result in the accumulation of starch.

The temporal separation of the nocturnal formation and storage of malic acid, and the diurnal decarboxylation and the accompanying accumulation of plastid starch, proceeds over the course of a 24-hour photoperiod. On this temporal basis, CAM consists of four metabolic phases that encompass the modulation of $C_4$ (PEPCase) and $C_3$ (Rubisco) carboxylation within the same cellular environment (Figure 1). The nocturnal uptake of atmospheric $CO_2$ via PEPCase to form $C_4$ acids and the diurnal decarboxylation of malic acid to elicit an elevated internal concentration of $CO_2$ are interspersed with transitional periods of net $CO_2$ uptake in early morning and in late afternoon, when both the PEPCase-and the Rubisco-mediated carboxylations contribute to the assimilation of $CO_2$. In this circadian rhythm, the proportion of $CO_2$ taken up via PEPCase at night or via Rubisco during the day is controlled by:

- stomatal function
- fluctuations in the content of organic acids and storage carbohydrates

- the activity of enzymes involved in the primary (PEPCase) and the secondary (Rubisco) carboxylations, and in the decarboxylation of organic acids (e.g., malic enzyme or PEP-carboxykinase)

In diurnal phase III, the closed stomata preclude gas exchange with the external atmosphere, but photosynthesis and other vital processes continue. At this stage, the internal levels of $CO_2$ fall precipitously due to an active Benson–Calvin cycle; however, this decline is offset by mitochondrial respiration. The closure of the stomata also leads to the accumulation of $O_2$ in leaf cells because the rate of photosynthetic $O_2$ evolution is higher than mitochondrial respiration and other $O_2$-consuming reactions. At this stage, the oxygenase activity of Rubisco activates the photorespiratory cycle by which the Benson–Calvin cycle recovers 75% of the carbon diverted to glycolate, with the remaining 25% released as $CO_2$ by the mitochondrial glycine decarboxylase complex. Thus, the closed stomata provide an adaptive advantage that minimizes water loss from the leaves by transpiration, and keeps the concentration of stromal $CO_2$ adequate for carbon assimilation. However, the concurrent accumulation of $O_2$ is a major adverse consequence because, as a powerful oxidant, $O_2$ reacts with other molecules and generates harmful free radicals (e.g., superoxide, hydroxyl ion) which may initiate uncontrolled adverse reactions on DNA, proteins, and lipids.

## Fructans

Although starch is the dominant carbohydrate reserve in many species, it may be almost absent in the storage organs of some plants. Fructans are less familiar reserve carbohydrates but occur in 15% of flowering species that mainly belong to the Asteraceae, Campanulaceae and Boraginaceae (dicots), and Poaceae and Liliaceae (monocots).

This type of polysaccharide, whose basic unit is fructose, was found in 1804 in a hot water extract from *Inula helenium* and later named "inulin." Fructans can replace starch as a reserve carbohydrate in some plants, though in many species both polysaccharides coexist. Moreover, the carbohydrate reserve substances vary among plant species and even within a single plant species throughout the seasons and under various climatic conditions. In contrast to starch—which is stored in plastids—fructans appear in vacuoles of different organs: for example, in bulbs (onion), tubers (Jerusalem artichoke, dahlia), taproots (chicory), leaves, and in stems (wheat, barley).

Chemically, fructans are distinguished on the basis of the glycosidic bond that links fructose residues to each other. The biosynthesis of these polysaccharides starts with the incorporation of a fructose moeity to one of the three primary hydroxyl groups of vacuolar sucrose—that is, carbon-1 and carbon-6 in the fructose moeity, and carbon-6 in the glucose moeity.

This reaction, catalyzed by the enzyme sucrose:sucrose 1-fructosyl transferase, implies a transglycosylation in which one molecule of sucrose transfers the fructosyl moeity to another. Thus, the addition of β-D-2,1-or β-D-2,6-linked fructofuranosyl units to the fructose moeity of sucrose yields 1-kestose and 6-kestose, respectively, while the addition of a β-D-2,6-linked fructofuranosyl unit to the glucose moeity of sucrose produces neo-kestose.

Further elongation of the ketose-type trisaccharide proceeds by the action of specific enzymes that transfer, also via transglycosylation, a fructosyl unit from a fructan molecule to positions 1 or 6 in the second fructose moeity of the trisaccharide. Thus, fructan:fructan fructosyl transferases catalyze the elongation of kestoses yielding the linear β-D-2,1-bonds of inulins, mainly in dicots, and the β-D-2,6-linkages in levans of monocots and bacteria.

$$\text{GFF (kestose)} + G(F)_m \text{ '! } G(F)_3 + G(F)_{m-1}$$

While up to 100,000 fructose moeities can be linked in a single molecule of bacterial fructan, fructosyl units range from 10 to 200 in plant counterparts. Fructans generally are present as a continuum of oligofructans that differ from each other in one fructose residue: $G(F)_3$, $G(F)_4$, $G(F)_5$, $G(F)_6$,..., $G(F)_n$. This feature enables fructans to participate in the regulation of the cellular osmotic potential.

In addition to supplying hexose units for demands of cellular energy, the breakdown of the endogenous polysaccharide elevates the concentration of lower members in the oligofructans series, quickly increasing in consequence the osmotic pressure.

Two types of fructan hydrolases cleave the glycosidic bond in fructans: exohydrolases that release terminal fructosyl residues while endohydrolases cleave at random within the fructan chains. Apparently, plants contain only the former type while bacteria and fungi contain both kinds of enzymes.

The recent finding of fructan exohydrolases in nonfructan plants not only suggests that these enzymes are additionally involved in other important functions—perhaps defence and signalling—but also highlights the necessity of functional characterization of these enzymes (Van den Ende et al. 2004).

## Plastidic Phosphate Translocators

Chloroplasts are metabolic powerhouses within leaf cells. They are not only the main site of carbon, nitrogen, and sulfur assimilation but also share metabolic pathways—such as the photorespiratory carbon cycle—with other plant cell compartments. A continuous exchange of metabolites and ions with the cytosol is necessary for playing these roles.

In all vascular plants, plastids are surrounded by an envelope that restricts the nonspecific diffusion of polar molecules. Thus, a pair of concentric membranes modulate the exchange of metabolites between the stromal interior and the cytosol. Although many pore-forming proteins that are substrate-specific were characterized in the outer membrane, the inner membrane is considered the main permeability barrier between the cytosol and the chloroplast. All transporters of the inner membrane are nuclear-encoded membrane proteins whose precursors are synthesized in the cytosol and post-translationally imported into the inner membrane. Chloroplast transporters collectively grouped as phosphate translocators catalyze a strict 1:1 counter-exchange (antiport) of orthophosphate with triose phosphates, pentose phosphates, and hexose phosphates.

The Arabidopsis genome contains six genes that encode four classes of functional plastidial phosphate translocators: single copies of the triose phosphate/phosphate translocator (TPT) and the phosphoenolpyruvate/phosphate translocators (PPT); and two copies each of the glucose 6-phosphate/phosphate translocators (GPT) and the xylulose 5-phosphate/phosphate translocator (XPT).

### *The Triose Phosphate/Phosphate Translocator and the Allocation of Carbon in Plants*

Triose phosphates, generated by the Calvin cycle at the expense of photosynthetically generated ATP and NADPH, flow to the cytosol across the chloroplast envelope. In turn, the orthophosphate released in the cytosol from biosynthetic processes is transported back into chloroplasts to replenish the ATP necessary for sustaining the assimilation of $CO_2$ and photosynthetic electron transport. Both the efflux of triose phosphates and the influx of orthophosphate are driven by a dimer composed of two identical subunits—the chloroplastic TPT. This membrane protein functions as an antiport system that exchanges phosphorylated—mainly three-, but also admits five-or six-carbon compounds—for orthophosphate. Hence, the chloroplast TPT exports the fixed carbon for the cytosolic synthesis of sucrose, the main product allocated to heterotrophic plant organs for further metabolism or conversion into storage products (e.g., starch, fructans). Therefore, the chloroplast TPT represents *the diurnal path for carbon export.*

Mutants defective in chloroplast TPT rely on different paths to transfer the photoassimilate from the chloroplast to the cytosol and, in so doing, compensate for the deficiency in TPT activity. Therefore, knockout mutations of this translocator increase the allocation of recently fixed carbon into transitory starch, followed by the degradation of the starch in the light and the subsequent export of neutral sugars to the cytosol. On the other hand, the simultaneous knockout of chloroplast TPT and starch synthesis drastically impairs plant growth.

### *The Phosphoenolpyruvate/Phosphate Translocator is Essential for the Chloroplast Metabolism*

PPT in the inner chloroplast membrane also links the stromal metabolism of all plants with the surrounding cytosol. $C_4$-plants of the NADP-malic enzyme type (e.g., maize) under active photosynthesis produce pyruvate in bundle sheath cells. This three-carbon compound returns to mesophyll cells for the conversion to phosphoenolpyruvate by the chloroplast-localized pyruvate-phosphate dikinase (Figure below). Subsequently, the phosphoenolpyruvate flows from the stroma to the cytosol for the primary carbon fixation via PEPCase. To ensure high rates of phosphoenolpyruvate efflux for the proper functioning of this cycle, the envelope of mesophyll chloroplasts bears the PPT, which concurrently drives the chloroplast uptake of orthophosphate to restore the lost phosphorus.

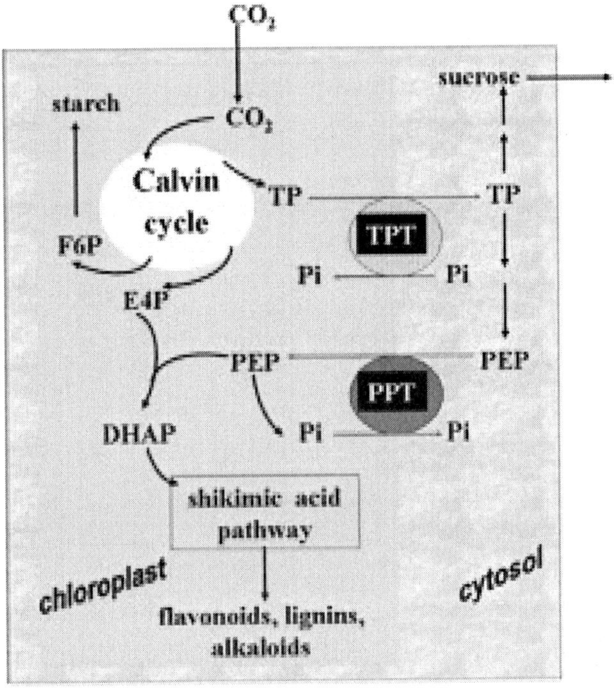


***Figure:*** *Mobilization of phosphoenolpyruvate in chloroplasts of* $C_3$ *plants. Under active photosynthesis, TPT furnish the cytosol with carbon skeletons for the biosynthesis of phosphoenolpyruvate. Upon the release of orthophosphate, the uptake of the cytosolic phosphoenolpyruvate, which combines in the stroma with erythrose 4-phosphate, triggers the formation of 3-deoxy-D-arabino-heptulosonic acid-7-phosphate (DHAP). DHAP is a scaffold for the biosynthesis of shikimic acid-5-phosphate, which in turn starts the pathway leading to the aromatic amino acids tyrosine and phenylalanine, the precursors of flavonoids, lignins, and alkaloids.*

On the other hand, CAM plants of the malic enzyme-type decarboxylate during the day the malate accumulated at night. Again, two successive processes account for the production of phosphoenolpyruvate, the primary acceptor of $CO_2$. First, the pyruvate formed in decarboxylation of the malate enters into the chloroplast stroma for its transformation to phosphoenolpyruvate catalyzed by the pyruvate-phosphate dikinase. Second, the phosphoenolpyruvate leaves the chloroplast in exchange for orthophosphate via the PPT. Knowledge on the function of the PPT in $C_3$ plants or nongreen tissues of $C_4$ plants is scarce. Apparently, the uptake—rather than the export—of phosphoenolpyruvate is the primary function of PPT in these tissues. Chloroplasts and most nongreen plastids are devoid of phosphoglyceromutase and enolase and, consequently, are unable to convert 3-phosphoglycerate to phosphoenolpyruvate. However, phosphoenolpyruvate combines with erythrose 4-phosphate, an intermediate of the pentose-phosphate pathway, for the synthesis of aromatic compounds through the shikimic acid pathway (Figure below). Thus, the provision of cytosolic phosphoenolpyruvate to the stroma via the PPT would be essential for the biosynthesis of flavonoids, lignins, alkaloids.

### *The Glucose 6-Phosphate/Phosphate Translocator Functions in Nongreen Tissues and is Essential for Plant Development*

The genome of Arabidopsis harbours two paralogous GPT genes, *AtGPT1* and *AtGPT2*, that, apparently, function in plastids of nongreen tissues for the import of glucose 6-phosphate. Two transgenic lines, bearing T-DNA insertions in the *GPT1* gene, exhibited not only impaired pollen and ovule development but also were lethal in the homozygous state. At variance, the disruption of the *GPT2* genes have no effect on growth and development. On the basis of drastic gametogenesis defects, it appears that the import of glucose 6-phosphate into nongreen plastids by the translocator AtGPT1 is crucial for the maturation of the pollen and the development of the female gametophyte.

### *The Xylulose 5-Phosphate/Phosphate Translocator*

The XPT constitutes the fourth subfamily of phosphate translocators that exhibits substrate specificity similar to the GPT but lacks the capacity to transport glucose 6-phosphate. In Arabidopsis, the lack of cytosolic transketolase and transaldolase halts the cytosolic oxidative pentose phosphate pathway at the stage of pentose phosphates (Kruger and von

Schaewen 2003). In this context, the putative function of the XPT would be to furnish the stromal pentose phosphate pathway with cytosolic five-carbon skeletons in the form of xylulose 5-phosphate, especially under conditions that require its removal from the cytosol.

## Photosynthesis: Physiological and Ecological Considerations

### *Working with Light*

Three light properties are especially important when working with light: amount, direction, and spectral quality. The first two parameters, amount and direction, are important with respect to the geometry of the part of the plant that intercepts the light. Is the plant part flat or cylindrical? For a flat leaf, a planar light sensor is the most appropriate, and the amount of energy that falls on a flat sensor of known area per unit time is quantified as irradiance. Units can be expressed in terms of energy, such as watts per square meter (W $m^{-2}$). Time (seconds) is contained within the term watt: 1 W = 1 joule (J)$s^{-1}$. The energy of a photon depends on its frequency, as expressed by Planck's law.

When considered as a wave, light has a wavelength and a frequency. Light can also be thought of as a stream of particles, photons, or quanta. In this case, units can be expressed in moles per square meter per second (mol $m^{-2}$ $s^{-1}$), where "moles" refers to the number of photons (1 mol of light = $6.02 \times 10^{23}$ photons, Avogadro's number). This measure is called photon irradiance.

Quanta and energy units can be interconverted, provided that the wavelength of the light is known. The energy of a photon is related to its wavelength as follows:

$$E = \frac{hc}{\lambda}$$

where *c* is the speed of light ($3 \times 10^8$ m $s^{-1}$), *h* is Planck's constant ($6.63 \times 10^{-34}$ J s), and λ is the wavelength of light, usually expressed in nm (1 nm = $10^{-9}$ m). We can solve for the *hë* part of the equation, and we obtain $1{,}988 \times 10^{-16}$, and write this equation as:

$$E = \frac{1,988 \times 10^{-16} J}{\lambda}$$

where λ is expressed in nanometers. From this equation we can see that a photon at 400 nm, which is in the blue region of the spectrum, has twice the energy of a photon at 800 nm, from the infrared region of the spectrum. A photon of 400 nm light contains $4.97 \times 10^{-19}$ J. On the other hand, the 800 nm photon contains $2.48 \times 10^{-19}$ J. Stated differently, the higher the wavelength of a photon, the lower its energy, as indicated by the larger denominator in the equation.

Now, suppose we have 3 μmol of 400 nm light falling on 1 $m^2$ every second, or a photon irradiance of 3 μmol $m^{-2}$ $s^{-1}$. This quantity is approximately the amount of blue light at 400 nm that strikes the surface of Earth on a sunny day. If we want to convert photon irradiance to energy irradiance, we must first convert micromoles to moles (1 μmol = $10^{-6}$ mol): (3 μmol $m^{-2}$ $s^{-1}$) × ($10^{-6}$ mol $\mu mol^{-1}$) = $3 \times 10^{-6}$ mol $m^{-2}$ $s^{-1}$ of photons, or quanta. Calculating the number of quanta from Avogadro's number gives: ($3 \times 10^{-6}$ mol quanta $m^{-2}$ $s^{-1}$) × ($6.02 \times 10^{23}$ quanta $mol^{-1}$) = $1.8 \times 10^{18}$ quanta $m^{-2}$ $s^{-1}$. As we already calculated, each photon, or quantum, of 400 nm light contains $4.97 \times 10^{-19}$ J. Thus ($4.97 \times 10^{-19}$ J $quantum^{-1}$) × ($1.8 \times 10^{18}$ quanta $m^{-2}$ $s^{-1}$) = 0.9 J $s^{-1}$ $m^{-2}$. Since 1 J $s^{-1}$ = 1 W, we have an irradiance of 0.9 W $m^{-2}$.

Direction of light. Turning our attention to the direction of light, light can strike a flat surface directly from above or it can strike the surface obliquely. When light deviates from perpendicular, irradiance is proportional to the cosine of the angle at which the light rays hit the sensor.

Thus, irradiance is maximal when light strikes a surface directly from above, and it decreases as light becomes more oblique—similar to the situation with a typical leaf. Sensors that correct for the angle of incidence of light are said to be *cosine corrected*. There are many examples in nature in which the light-intercepting object is not flat (e.g., complex shoots, whole plants, chloroplasts). In addition, in some situations light can come from many directions simultaneously (e.g., direct light from the sun plus the light that is reflected upward from sand, soil, or snow). In these situations it makes more sense to measure light with a spherical sensor that measures light omnidirectionally (from all directions). When the amount of light is measured by this omnidirectional measurement, the type of measurement is called fluence rate (Rupert and Letarjet 1978), and the measured amount of light can be expressed in watts per square meter (W $m^{-2}$) or moles per square meter per second (mol $m^{-2}$ $s^{-1}$). It is clear from the units whether light is being measured as energy (W) or as photons (mol).

In contrast to a flat sensor, a spherical sensor is equally sensitive to light from all directions. Depending on whether the light is collimated (rays are parallel) or diffuse (rays travel in random directions), values for fluence rate versus irradiance can be quite different from one another. They are equivalent only under special conditions.

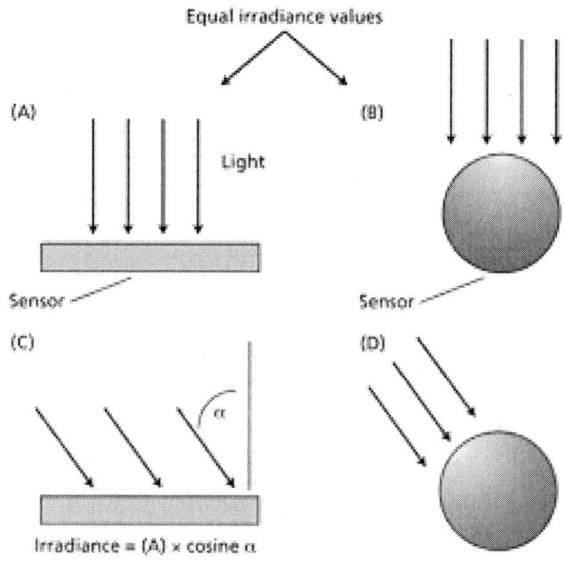


***Figure:*** *Irradiance and fluence rate. Equivalent amounts of collimated light strike a flat irradiance-type sensor (A) and a spherical sensor (B) that measure fluence rate. With collimated light, A and B will give the same light readings. When the light direction is changed 45°, the spherical sensor (D) will measure the same quantity as in B. In contrast, the flat irradiance sensor (C) will measure an amount equivalent to the irradiance in A multiplied by cosine of the angle α in C. (After Björn and Vogelmann 1994.)*

The flat sensor measurement of photosynthetically active radiation (PAR, 400 to 700 nm) may also be expressed on the basis of energy (W $m^{-2}$) or quanta (mol $m^{-2}$ $s^{-1}$) (McCree 1981). It is important to note that PAR is an irradiance-type measurement. In research on photosynthesis, when PAR is expressed on a quantum basis, it is often given the special term photosynthetic photon flux density (PPFD). However, it has been suggested that the term "density" be discontinued because within the International System of Units (Système Internationale d'Unités, or SI units) "density" can mean area or volume. Moreover, area is contained within the term flux. PPFD has in some cases been shortened to PPF, but it is not clear whether this abbreviation represents an irradiance-type or a spherical measurement.

When light moves through water such as lakes or oceans, it scatters in such a way that its *direction* of travel is randomized. As a result, aquatic photosynthetic organisms can receive light simultaneously from above, below, and the sides. In oceanography and limnology, the light that travels from the water surface toward the bottom is called *downwelling irradiance*, and the light that travels in the reverse direction is called *upwelling irradiance.*

Before the advent of specialized light meters suitable for plant physiology, light was measured in lux or foot-candles. These measurements are based on the perception of light by the human eye, which is maximally sensitive to light within the green region of the spectrum, at 555 nm. Sensitivity falls off on both sides of this wavelength and approaches zero within the blue and red, wavelength regions that are important for photosynthesis. Thus, lux and foot-candles are of little use in plant physiology. These units can be converted to the units of irradiance and fluence rate, but only if detailed knowledge about the spectral quality of the light is available, something that can be obtained only by use of an instrument called a *spectroradiometer*, which measures the amount of light at each wavelength. In summary, when choosing how to quantify light, it is important to match sensor geometry and spectral response with that of the plant. Flat, cosine-corrected sensors are ideally suited to measure the amount of light that strikes the surface of a leaf; spherical sensors are more appropriate in other situations, such as when studying a chloroplast suspension or a branch from a tree.

How much light is there on a sunny day and what is the relationship between PAR irradiance and PAR fluence rate? Under direct sunlight, PAR irradiance and fluence rate are both about 2000 μmol $m^{-2}$ $s^{-1}$, though higher values can be measured at high altitudes. The corresponding value in energy units is about 400 W $m^{-2}$. When light is completely diffuse, irradiance is only 0.25 times the fluence rate.

## Heat Dissipation from Leaves: The Bowen Ratio

The heat load on a leaf exposed to full sunlight is very high. In fact, a leaf with an effective thickness of water of 300 μm would warm up by 100°C every minute if all available solar energy were absorbed and no heat was lost. However, this enormous heat load is dissipated by the emission of long-wave radiation, by sensible (or perceptible) heat loss, and by evaporative (or latent) heat loss.

Air circulation around the leaf removes heat from the leaf surfaces if the temperature of the leaf is higher than that of the air; this phenomenon is called sensible heat loss.

Evaporative heat loss occurs because the evaporation of water requires energy. Thus, as water evaporates from a leaf it withdraws heat from the leaf and cools it. The human body is cooled by the same principle, through perspiration.

Sensible heat loss and evaporative heat loss are the most important processes in the regulation of leaf temperature, and the ratio of the two is called the Bowen ratio:

$$\text{Bowen ratio} = \frac{\text{sensible heat loss}}{\text{evaporative heat loss}}$$

This concept was developed by Ira S. Bowen (1898–1978), an American astrophysicist. When the evaporation rate is low, because water supply is limited, the Bowen ratio tends to be high. Thus, the Bowen ratio is about 10 for deserts, 2-6 for semi-arid regions, 0.4 to 0.8 for temperate forests and grasslands, 0.2 for tropical rain forests and 0.1 for tropical oceans

In well-watered crops, transpiration, and hence water evaporation from the leaf, is high, so the Bowen ratio is low.

On the other hand, in some cacti, stomata closure prevents evaporative cooling; all the heat is dissipated by sensible heat loss, and the Bowen ratio is infinite. Plants with very high Bowen ratios conserve water but have to endure very high leaf temperatures in order to maintain a sufficient temperature gradient between the leaf and the air. Slow growth is usually correlated with these adaptations.

Very low Bowen ratios can be measured in a lawn on a relatively still day. In these conditions there is no sensible heat loss, because the air around the leaf is at the same temperature as the leaf; the Bowen ratio therefore approaches zero, and heat dissipation is due mostly to evaporative heat loss.

In other cases, such as cotton leaves in the afternoon, water loss from stomatal transpiration cools the leaf below the air temperature by evaporative heat loss. In that case, there is sensible heat gain rather than heat loss, and the Bowen ratio becomes negative.

One can calculate the evapotranspiration rate for an entire canopy using measurements of the Bowen ratio, net incident radiation, the heat loss from the soil, and the gradients in temperature and water vapour concentration above the canopy.

## The Geographic Distributions of $C_3$ and $C_4$ Plants

Among the 15,000+ species with $C_4$ photosynthesis, it is most common in grasses and sedges, less common in herbs and shrubs, and not found in trees (with a single Hawaiian tree exception, *Euphorbia forbesii*). Climate is a major factor influencing the natural distributions of $C_3$ plants and $C_4$ grasses. Here, we talk about the two most important climate parameters influencing plant growth: water and temperature.

Clearly plants will not grow in the absence of water, so the important factor influencing photosynthetic pathway distribution becomes the temperature during the growing period. Based on many systematic surveys of the natural vegetation across the globe that have been accumulating over time, a clear picture is emerging. $C_4$ taxa are found in warm to temperate environments and are uncommon in cool to cold climates. Below we construct a global map that describes the general abundances of $C_3$ and $C_4$ taxa on different continents.

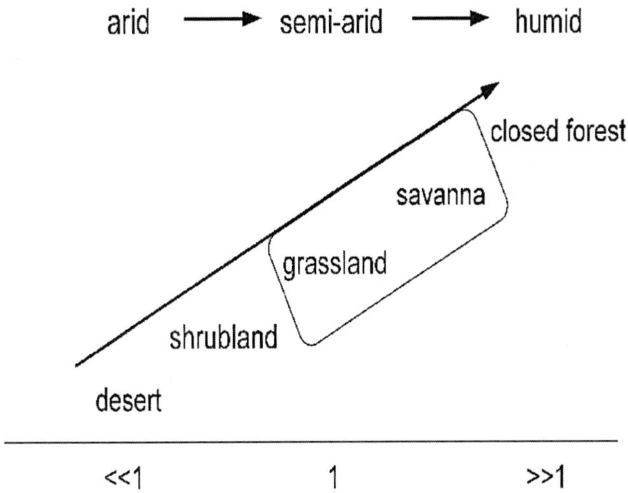


***Figure:*** *A plot of the vegetation types that are expected to occur along a climate gradient estimated by the ratio of the incoming precipitation to the potential evaporation rate. Courtesy of Ehleringer, unpublished figure.*

Notice that $C_4$ taxa are not very common in tropical regions (0–20° latitude), because dense tropical forests tend to shade out $C_4$ grasses. The $C_4$ taxa are most common away from the tropics in the savanna and steppe regions; their abundances tends to diminish south of the desert zones (generally 30–40° latitude). Climate can be presented along an axis of an increasing of precipitation relative to evaporative water loss. At the upper, wetter end of this gradient, we would find the forests. At the drier end, where precipitation is low relative to evaporative demand, we see the deserts. In the figure below, we see that these different vegetation zones reflect variations in the ratios of precipitation-to-evaporation. $C_4$ plants are most common in this semi-arid region.

In deserts with wet monsoon summers, such as in central Africa, the sub-Indian continent, and western North America, $C_4$ taxa are abundant. In deserts that receive winter rainfall, such as deserts in Asia and North Africa, there are few $C_4$ taxa. In North America, this regional influence of summer precipitation is most visible in the southwestern United States where there are large differences in precipitation associated with the Arizona monsoon boundary as shown below.

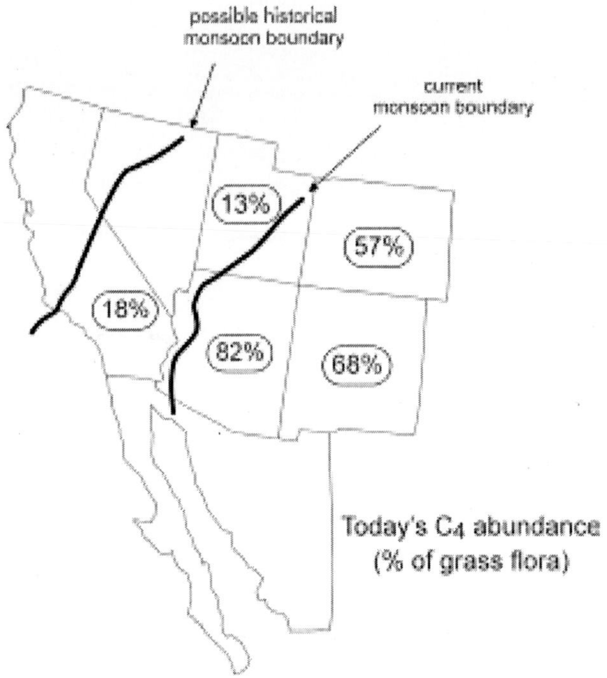


***Figure:*** *Calculations of the percentage of the grass species that have* $C_4$ *photosynthesis in different states of the western United States. Note that the presences of summer rain influences* $C_4$ *abundance.* $C_4$ *plant tend to occur in warm, summer wet regions. Courtesty of Ehleringer, Cerling, and Dearing (2005).*

Human activities and disturbances by animals will influence the distribution of $C_3$ and $C_4$ taxa within savannas and grasslands. This reflects the role of disturbances, such as grazing and fires, on the importance of trees on the landscape.

## Projected Future Increases in Atmospheric $CO_2$

Human use of fossil fuels (coal, oil, and natural gas) continues to increase as growing human populations demand more energy for transportation, heating, and manufacturing. We measure atmospheric $CO_2$ in units of ppm or parts per million.

The rate of atmospheric increase in $CO_2$ is about 3 ppm per year. Note the cyclic nature of the atmospheric $CO_2$ data, in which one oscillation cycle is exactly one year. This annual pattern reflects changes in the balance of photosynthesis (decreases atmospheric $CO_2$) and respiration (increases atmospheric $CO_2$) at a location over the course of the year. Atmospheric $CO_2$ tends to decrease in the spring and summer, when photosynthesis rates within an ecosystem exceed respiration rates. In contrast, atmospheric $CO_2$ tends to increase in the fall and winter when respiration rates exceed photosynthesis. In 2007 atmospheric $CO_2$ reached an average value of 384 ppm and is expected to reach 400 ppm before 2015.

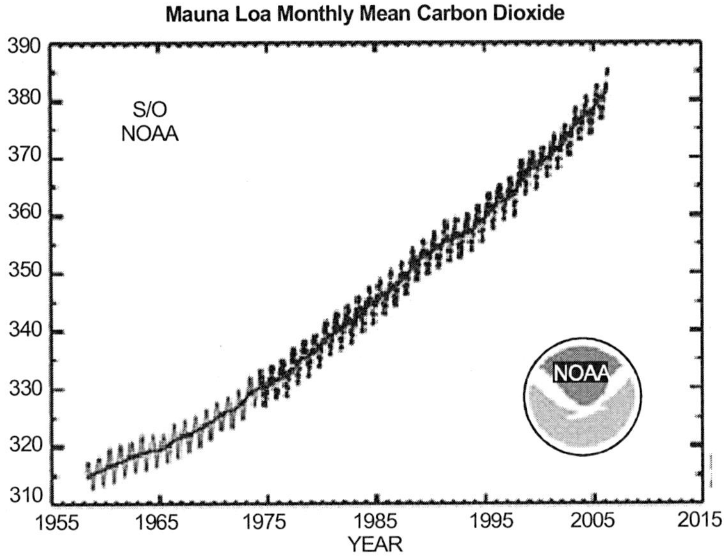


*Figure: A high precision record of the atmospheric carbon dioxide levels measured on Manua Loa, Hawaii.*

Economists have good estimates of the rate of $CO_2$ emission globally. The U.S., European countries, China, Japan, and India are the largest sources of fossil fuel emissions.

One surprising fact is that the observed rate of atmospheric $CO_2$ increase is actually less than the observed rate of atmospheric $CO_2$ increase.

This is because plants on land and algae in the ocean are currently able to take up about one-half of fossil fuel emissions through enhanced photosynthesis. Scientists study how plants and ecosystems respond to elevated $CO_2$ using an experimental field approach called a Free Air $CO_2$ Enrichment (FACE)

Experiment. In a FACE experiment, pipes inject $CO_2$ into the interior of a ringed area containing a complete ecosystem as shown below.

These FACE research facilities give scientists an opportunity to understand how different plant biochemical, physiological, and growth processes within the ecosystem will respond as a result of long-term exposure to elevated $CO_2$ levels. Since biomass production involves so much more than simply increased photosynthesis (i.e., mineral nutrients are required as well), it is doubtful that plant growth can be sustained in a linear, proportional fashion as atmospheric $CO_2$ levels continue to increase.

The FACE studies are designed to address the question of how ecosystems will respond to future atmospheric $CO_2$ environments and whether the growth response level off at some future $CO_2$ level. Global warming and changes in climate are anticipated effects of a rapidly increasing $CO_2$ levels. These are, of course, but two of the many reasons why scientists and others are concerned about the consequences of elevated atmospheric $CO_2$. Just how much the atmospheric $CO_2$ will increase is unknown. Below are estimates of the ranges, based on two plausible scenarios.

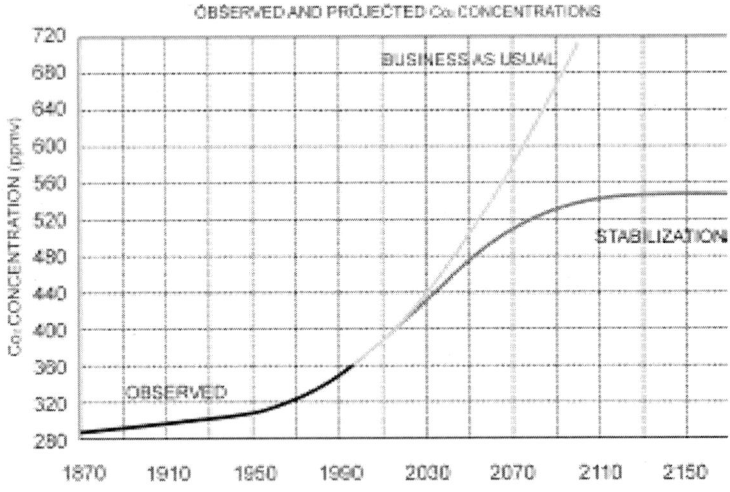


***Figure:*** *Projected changes in the atmospheric $CO_2$ concentrations under constrasting emission-control projections.*

In one scenario, titled "business as usual" atmospheric $CO_2$ levels are projected to reach 700 ppm by the end of this century. On the other hand, an aggressive global effort to curb $CO_2$ emission might result in an atmospheric $CO_2$ stabilization of 550 ppm.

## Using Carbon Isotopes to Detect Adulteration in Foods

Carbon isotope ratios ($\delta^{13}C$) of $C_4$ plants such as corn and sugar cane are –10 to –12 ‰, in contrast to $\delta^{13}C$ values of $C_3$ foods, which are typically –26 to –28‰

The naturally occurring variations in carbon isotope ratios within the $C_3$ of $C_4$ ranges occur because environmental factors influence carbon isotope ratios. However, the differences in carbon isotope ratios among the two photosynthetic pathways are large enough to ensure that it is possible to distinguish between whether or not a food substance is based on $C_3$ of $C_4$ photosynthesis. Adulteration is the substitution of a cheaper food substance or compound with another. Carbon isotopes are frequently used to detect the substitution of $C_4$ sugars for $C_3$ food products, such as the introduction of sugar cane into honey to increase yield. One easy way to detect this adulteration is to examine the jams and jellies that are found in restaurants. Below are the results of a survey conducted as a science project by Jeff Ehleringer when he was a young man in junior high school. Jeff collected jams and jellies every time he went into a restaurant. He had the opportunity to collect jams and jellies from across the western United States and in parts of Europe as well.

The fruits that go into producing jams and jellies are all $C_3$ fruits (grapes, strawberries, raspberries, orange, apricots, peaches, etc.) One subgroup of the jams and jellies appear to have carbon isotope ratios similar to the stable isotope ratios of $C_3$ plants, from which they were supposed to have been derived. These were all samples that Jeff had collected in Europe, where there are authenticity laws regulating what materials can go into making a processed food such as a jelly. In contrast, the vast majority of the jams and jellies had carbon isotope ratios consistent with those of $C_4$ plants. Yet, fruits are not from $C_4$ plants but instead are produced from $C_3$ plants. These are the adulterated jellies, most likely produced largely from sugar syrup and food flavouring with very little food used in the process. These substitutions are legal in the U.S., because here, we do not have authenticity laws with respect to food production. A third small subgroup consists of jams that were made of 50/50 $C_3/C_4$ food sources.

It turns out that these were the more expensive jams and jellies. There are many other food products that are tested for their carbon isotope ratios to detect adulteration. Common among these are fruit juices, vanilla (vanilla beans come from CAM plants), beer (wheat and barley are $C_3$), and alcoholic beverages.

## Reconstruction of the Expansion of $C_4$ Taxa

The rate of photorespiration in a leaf increases because RuBP carboxylase is more likely to react with $O_2$ instead of $CO_2$ as temperature increases and/or $CO_2$ decreases. The result of an increase in photorespiration is a decrease in photosynthetic quantum efficiency. This leads to environmental conditions where $C_4$ plants are favoured over $C_3$ and vice versa.

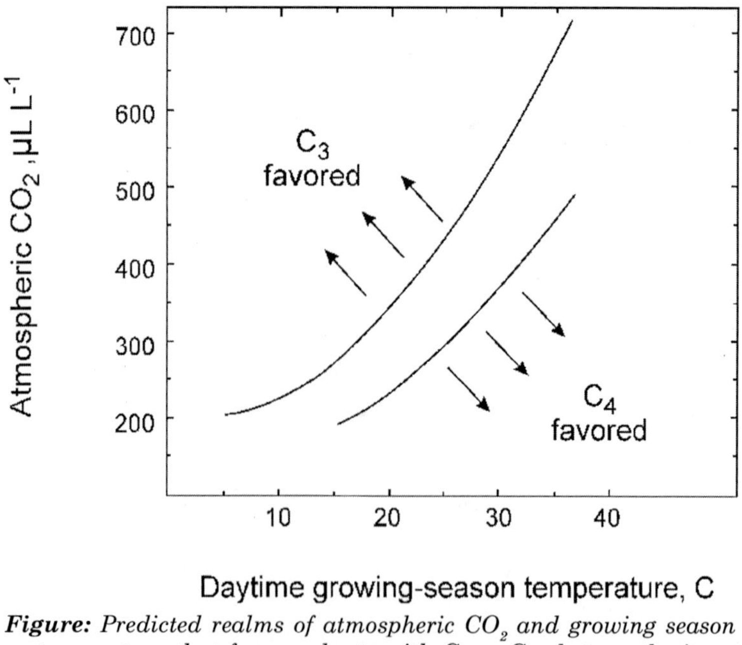


***Figure:*** *Predicted realms of atmospheric $CO_2$ and growing season temperature that favour plants with $C_3$ or $C_4$ photosynthesis.*

This model leads to predictions that the abundances of $C_3$ and $C_4$ plants should vary spatially and as a function of atmospheric $CO_2$ history. The secret to evaluating this model is to go back and look at plant fossils over time and to relate photosynthesis pathway with prehistoric atmospheric $CO_2$. This is a difficult task. Reconstruction of the expansion of $C_4$ plants over the past 10–15 million years can be a challenge, because individual plants are not well recorded in the fossil record in many locations. Most plant fossils are thought to have formed as a result of submersion into an anaerobic, aquatic environment. The vast majority of plants do not become fossils under these conditions, but instead are decomposed by microbial activities leading to diminished numbers of fossils that we can use to reconstruct the expansion of $C_4$ plants over time. However, there are good proxies. Carbon isotope ratios ($\delta^{13}C$) of animal tissues reflect the foods that they ate. For modern animals, it is possible to choose from among, hair, muscle, bone collagen, and tooth enamel for isotope analyses in order to determine the proportions of $C_3$ and $C_4$ food sources in their diets. Hair provides a sequential record of the animal's diet during the time of hair production. Muscle and bone collagen analyses provide an integrated measure of the animal's current diet as shown in the figure below.

The $\delta^{13}$13C of animal teeth faithfully record the carbon isotope ratios of food sources during the period of tooth development. The carbon in tooth enamel is actually carbonate that has condensed from $CO_2$ that was produced as a byproduct of metabolizing food. After an animal has died and decomposition has occurred, about the only remaining tissue that remains and preserves the $C_3/C_4$ dietary signals is the enamel in the animal's teeth. For many herbivores, such as cows and horses, tooth growth is continuous and provides a longer-term dietary record than for animals with a fixed tooth growth period (such as humans). As teeth are preserved for millions of years, $\delta^{13}$ of tooth enamel can be used to reconstruct the abundances of $C_3$ and $C_4$ plants eaten by mammalian grazers over extended time periods. In the graphic below, we plot the frequency distributions of modern animal teeth with that observed for $C_3$ and $C_4$ plants:

Note that there is an "å" offset of 14.1‰. This is the isotope enrichment associated with $CO_2$ condensing to form carbonate.

Now we have a tool—the carbon isotope ratio of animal tooth enamel will record the abundance of $C_3$ versus $C_4$ food sources.

Until 8 million years ago, $C_4$ plants were not an important part of ecosystems. Then, between 6 and 9 million years ago, $C_4$ plants became important parts of many ecosystems, particularly at tropical and semi-tropical latitudes. These results are consistent with model predictions above. $C_4$ photosynthesis is predicted not to occur on Earth until the atmospheric $CO_2$ level falls below a critical threshold value. Even then it should appear first in those locations with the warmest temperatures during the growing season.

There is also extensive evidence presently to show that fluctuations in the atmospheric $CO_2$ concentration values between glacial (180 ppm) and interglacial (280 ppm) periods were large to influence the local abundances of $C_3/C_4$ plants. Shown below are the time series analyses of carbon in bog in tropical Africa. At about 10,0000 to 12,000 years ago, the vegetation surrounding these bogs switched from being dominated by $C_4$ plants to being dominated by $C_3$ plants. This shift is evident by the dramatic change is carbon isotope ratios of materials feeding into the bog, especially at the time that the glacial period ended and we entered our current inter-glacial period.

Together these historical pieces of information paint an interesting history. The $C_3$ photosynthetic pathway dominated for much of Earth's history. And it is only relatively recently in Earth's (6–8 million years ago) history that we have seen an expansion of $C_4$-dominated ecosystems. What the future holds is unclear, since anthropogenic burning of fossil fuels is rapidly increasing in atmospheric $CO_2$ to levels far exceeding those observed on Earth over the past 1 million years.

# Chapter 4

# Cells and Viruses

## Cellular Organization

Cells are the smallest structural unit of living organisms, capable of maintaining life and reproducing. Viruses are not cells because they cannot maintain life and reproduce by themselves. Although a nerve cell looks entirely different from a red blood cell, their organizations are essentially the same. Even plant cells and animal cells share significant similarity in the overall organization.

### *Classification of Cells and Organisms*

All cells are divided into two types: prokaryotic cells and eukaryotic cells.

The prokaryotic cell does not have a nucleus.

The eukaryotic cell contains a nucleus.

Eukaryotes are the organisms made up of eukaryotic cells. They include protista, fungi, animals and plants. Prokaryotes include archaebacteria and eubacteria. They are single-cell organisms. More recently, "archaebacteria" have been placed in a category outside "bacteria", because they are quite different from the ordinary bacteria. According to the new classification, prokaryotes are divided into archaea and bacteria, where "archaea" is equivalent to "archaebacteria", and "bacteria" is the same as "eubacteria". Archaea live in extreme environments. They may be organized into three groups:

> *Methanogens live in anaerobic environment such as swamps. They produce methane and cannot tolerate exposure to oxygen. Extreme halophiles*

*live in very high concentrations of salt (NaCl), e.g., the Dead Sea and the Great Salt Lake. Extreme thermophiles live in hot, sulfur rich and low pH environment, such as hot springs, geysers and fumaroles in the Yellowstone National Park.*

### Basic Cellular Components

All cells contain cytoplasm, plasma membrane, and DNA.

Cytoplasm is the viscous contents of a cell, including proteins, ribosomes, metabolites and ions. Ribosomes are the sites of protein synthesis. Plasma membrane is the cell membrane surrounding cytoplasm. It consists of phospholipid bilayer, associated proteins and carbohydrates. Phospholipid bilayer is also the basic constituent of other biomembranes. DNA (deoxyribonucleic acid) is the genetic material. An eukaryotic cell contains several DNA molecules, located in the nucleus and mitochondria which are membrane-bound organelles. A prokaryotic cell contains a single DNA molecule, which has no specific boundary with the cytoplasm.

## Prokaryotic Cells

A prokaryotic cell consists of DNA, cytoplasm, and a surface structure which includes the plasma membrane and some of the following components:

- Cell wall
- Capsule
- Slime
- Flagella
- Fimbriae/Pili.

In bacteria, the cell wall contains a unique structure called peptidoglycan. Archaea do not possess peptidoglycan, but some archaea may contain pseudopeptidoglycan, which is composed of N-acetyltalosaminuronic acid, instead of N-acetylmuramic in peptidoglycan. Because of this structural difference, archaea are resistant to many cell wall antibiotics. Bacteria may be divided into two groups, on the basis of their cell wall structures and the response to Gram stain: Gram-negative and Gram-positive.

In Gram-negative bacteria, the cell wall is composed of three layers: (1) the periplasmic space which is an open area located outside the plasma membrane, (2) a thin layer of peptidoglycan external to the periplasmic space, and (3) the outer membrane surrounding the peptidoglycan. The Gram-positive bacteria do not have the periplasmic space and outer membrane, but have a thicker peptidoglycan layer. As a result, they are quite sensitive to lysozyme and penicillin. Capsule and slime are the hydrophilic gel surrounding the cell wall in most bacteria. The capsule is more closely associated with the cell than the slime. Flagella are long, rigid protein rods, facilitating the movement of motile bacteria. Fimbriae and pili are short hair-like structures used to attach other cells. They are essential for infecting other organisms.

### *Spores*

The spore is a small, often unicellular, reproductive unit of plants, algae, fungi, protozoa, and bacteria. Bacterial spores have thick walls which can withstand varying temperatures, humidity, and other unfavorable conditions.

## Eukaryotic Cells

The eukaryotic cell contains organelles, which are defined as membrane-bound structures such as nucleus, mitochondria, chloroplasts, endoplasmic reticulum (ER), Golgi apparatus, lysosomes, vacuoles, peroxisomes, etc. Prokaryotic cells do not have organelles. For animal cells, the cell surface consists of the plasma membrane only, but plant cells have an additional layer called cell wall, which is made up of cellulose and other polymers. The nucleus is the largest organelle in an eukaryotic cell. It is not part of the cytoplasm. By definition, cytoplasm is everything inside the plasma membrane except the nucleus. Under microscope, the nucleus shows two distinct areas. The darker area is called nucleolus, and the lighter area is known as nucleoplasm.

## Biomembranes

All biological membranes, including plasma membranes and all organelle membranes, contain lipids and proteins. The lipids found in biomembranes are mainly phospholipids and cholesterol. In the plasma membrane and some of organelle membranes, proteins and phospholipids are attached to carbohydrates, forming glycoproteins and glycolipids, respectively.

## Phospholipids

A phospholipid molecule consists of a hydrophilic polar head group and a hydrophobic tail. The polar head group contains one or more phosphate groups. The hydrophobic tail is made up of two fatty acyl chains. When many phospholipid molecules are placed in water, their hydrophilic heads tend to face water and the hydrophobic tails are forced to stick together, forming a bilayer.

### *Polar Head Groups*

Most phospholipid head groups belong to phosphoglycerides, which contain glycerol joining the head and the tail. Examples of phosphoglycerides include phosphatidylcholine, phospha tidylserine, phosphatidylethanolamine, phosphatidyl inositol, etc.

### *Cholesterol and Steroids*

Cholesterol is absent from most prokaryotic cells, but abundant in the plasma membrane of mammalian cells. It is used as a precursor to generate other important steroids.

Cholesterol plays a central role in atherosclerosis - a disorder that may cause heart attack or stroke. It is also involved in Alzheimer's disease.

### *Glycoproteins and Glycolipids*

Glycoproteins are the proteins covalently attached to carbohydrates such as glucose, galactose, lactose, fucose, sialic acid, N-acetylglucosamine, N-acetylgalactosamine, etc.

Glycolipids are carbohydrate-attached lipids. Their role is to provide energy and also serve as markers for cellular recognition.The antigens which determine blood types belong to glycoproteins and glycolipids (more info).

### *Blood Group Antigens*

The antigens which determine blood types belong to glycoproteins and glycolipids. There are three types of blood-group antigens: O, A, and B. They differ only slightly in the composition of carbohydrates.

## The Nucleus

The cell nucleus consists of nuclear envelope, nucleolus and nucleoplasm. Most chromosomes are located in the nucleoplasm, but portions of several chromosomes containing clusters of rRNA genes may get together in the nucleolus, forming the nucleolar organizing region. The major role of the nucleolus is to produce rRNA.

### *Introduction to Chromosomes*

Chromosomes are the structures that hold DNA molecules. One chromosome contains a DNA molecule. Each chromosome has a p and q arm; p is the shorter arm and q is the longer arm. The arms are separated by a pinched region called the centromere.

In order for chromosomes to be seen with a microscope, they need to be stained. Once stained, the chromosomes look like strings with light and dark "bands" and their picture can be taken. The picture, or chromosome map, is called a karyotype.

The germ cell (sperm or egg) of a human being contains 23 chromosomes, labeled from 1 to 22 and either X or Y. The somatic cells (cells other than germ cells) of a normal person has 46 chromosomes. For other species, the chromosome number varies from 1 to 1260.

A human somatic cell contains two chromosomes that determine the sex of a person. The two sex chromosomes are XY in male and XX in female. The gene (SRY) that is important for testis formation is located on chromosome Y. It is possible that a person with testis still exhibits female characteristics.

### *Chromosomes and Karyotype*

In a non-dividing cell, chromosomes are not visible by light microscopy, because chromatin spreads throughout the nucleus. During the metaphase of cell division, the chromatin condenses and becomes visible as chromosomes. At this time, each chromosome has been duplicated. A chromosome becomes two sister chromatids attached at the centromere.

### *Chromosome Banding*

To see chromosomes by microscope, they are normally treated with chemical dyes, such as Giemsa. The chromosome will appear as a series of alternate dark and light bands. If Giemsa is used, the dark band is called G-band or G-positive band, and the light band is named G-negative band. Similar banding patterns can be observed by using another dye, Quinacrine. However, if chromosomes were treated in a hot alkaline solution before staining with Giemsa, a reverse pattern will be observed, namely, the original dark band will become light band, and vice versa. For this reason, the G-negative band is also known as the R-band.

Chromosome bands are named as follows. Each chromosome consists of two arms separated by the centromere. The long arm and short arm are labeled q (for queue) and p (for petit), respectively.

At the lowest resolution, only a few major bands can be distinguished, which are labeled q1, q2, q3; p1, p2, p3, etc., counting from the centromere. Higher resolution reveals sub-bands, labeled q11, q12, q13, etc. Sub-sub-bands identified by even higher resolution are labeled q11.1, q11.2, q11.3, etc. Traditionally, the short arm (p) is displayed on top of the long arm (q).

### *Karyotype*

Karyotype is the representation of entire metaphase chromosomes in a cell, arranged in order of size.

### *Chromosome Numbers*

A germ cell (sperm or egg) contains only one set of chromosomes. It belongs to haploid, represented as 1n. Somatic cells (cells other than germ cells) of sexually reproducing organisms are diploid, denoted by 2n. In humans, the haploid chromosome number is 23, but the diploid chromosome number is 46.

### *Extremes*

Smallest number: The female of a subspecies of the ant, *Myrmecia pilosula*, has one pair of chromosomes per cell. Its male has only one chromosome in each cell.

Largest number: In the fern family of plants, the species *Ophioglossum reticulatum* has about 630 pairs of chromosomes, or 1260 chromosomes per cell.

### Sex Chromosomes

Sex chromosomes determine the sex of an organism. A human somatic cell has two sex chromosomes: XY in male and XX in female. A human germ cell has one sex chromosome: X or Y in a sperm and X in an egg. When an X-sperm is combined with an egg, the resulting zygote (fertilized egg) will contain two X chromosomes. A person developed from the XX-zygote will have the characteristics of a female. Combination of a Y-sperm and an egg will produce a male.

### The SRY Gene

Usually, a woman has two X chromosomes (XX) and a man one X and one Y (XY). However, both male and female characteristics can sometimes be found in one individual, and it is possible to have XY women and XX men. Analysis of such individuals has revealed some of the molecules involved in sex determination, including one called SRY, which is important for testis formation. SRY (which stands for sex-determining region Y gene) is found on the Y chromosome.

In the cell, it binds to other DNA and in doing so distorts it dramatically out of shape. This alters the properties of the DNA and likely alters the expression of a number of genes, leading to testis formation.

Most XX men who lack a Y chromosome do still have a copy of the SRY gene on one of their X chromosomes (moved there by chromosomal translocation). This copy accounts for their maleness. However, because the remainder of the Y chromosome is missing they frequently do not develop secondary sexual characteristics in the usual way.

## Cytoplasmic Organelles

By definition, organelles are the membrane-bound structures in a cell. The nucleus is an example. Other organelles are located in the cytoplasm such as mitochondria, chloroplasts, endoplasmic reticulum, Golgi apparatus, peroxisomes, lysosomes, vacuoles and glyoxisomes.

### Mitochondria

An eukaryotic cell contains many mitochondria, occupying up to a quarter of the cytoplasmic volume. The size of a mitochondrion is about 1.5-2 mm in length, 0.5-1 mm in diameter, approximately the same as *E. coli*. It has two membranes: outer membrane and inner membrane. Mitochondria also have their own DNA (represented as mtDNA), which encodes some of the proteins and RNAs in mitochondria. However, most proteins operating in mitochondria still originate from nuclear DNA.

The major role of mitochondria is to produce ATP (adenosine triphosphate), which carries high energy to power most cellular processes. Such energy is stored in the phosphoanhydride bonds of ATP. During ATP hydrolysis, the bond is broken, releasing 7.3 kcal/mole of energy. Many cellular processes can utilize the released energy by coupling with the ATP hydrolysis.

In animal cells, the major sources for the synthesis of ATP are fatty acids and glucose. Oxidation of an 18-carbon fatty acid can make 146 ATP molecules. By contrast, oxidation of one glucose molecule (6 carbons) can generate only 36 ATP molecules.

The generation of ATP involves a series of electron transport. Inevitably, electrons may leak from the electron transport chain, producing free radicals. This has been suggested to be the major mechanism involved in the aging process.

### Chloroplasts

Like mitochondria, a chloroplast also contains both outer and inner membranes on its surface. Inside the chloroplast, there are many thylakoids, each is enclosed by a membrane. Chlorophylls are located on the thylakoid membrane to absorb light for photosynthesis.

In the first step of photosynthesis, light energy is used to split water into hydrogen ions and oxygen molecules. The generated hydrogen ions will create a concentration gradient across the thylakoid membrane. Movement of hydrogen ions through the membrane is coupled to ATP synthesis. The overall reactions can be written as

### Endoplasmic Reticulum

Endoplasmic reticulum (ER) can be divided into rough ER and smooth ER. The major role of rough ER is to process the newly synthesized peptides from ribosomes. Therefore, the surface of rough ER is usually associated with ribosomes and thus appears "rough". Smooth ER is involved in the synthesis and metabolism of lipids. Hepatocytes are abundant in smooth ER.

### *Golgi Apparatus*

Golgi apparatus is a major site for sorting and modifications of proteins and lipids. After proteins are sorted at rough ER, they are enclosed in transport vesicles and carried to the Golgi apparatus. Some proteins could be modified into glycoproteins and then transported to other destinations.

### *Peroxisomes*

Peroxisomes contain enzymes for degrading amino acids and fatty acids. These reactions produce harmful hydrogen peroxide. Hence, peroxisomes also contain catalase to convert hydrogen peroxide into water and oxygen:

### *Lysosomes*

The major function of lysosomes is to degrade various macromolecules in the cell. They contain nuclease for degrading DNA and RNA, protease for degrading proteins and other enzymes for degrading polysacchrides and lipids. Lysosomes exist only in animal cells. Although plant cells do not have lysosomes, their vacuoles are also capable of degrading macromolecules..

### *Vacuoles*

Vacuoles store small molecules such as water, ions, sucrose and amino acids. They can also hold waste products which will be slowly degraded. They typically occupy more than 30% of the cell volume, but may expand up to 90%.

### *Glyoxisomes*

Glyoxisomes are found mainly in plant seeds. Their major function is to convert fatty acids into acetyl CoA for the glyoxylate cycle where two acetyl-CoA molecules are converted to a 4-carbon dicarboxylic acid. Peroxisomes and glyoxisomes are also called microbodies.

## Viruses

Viruses are the smallest organisms, with diameters ranging from 20 nm to 300 nm (1 nm = $10^{-9}$ meter). Viruses are not cells.

They consist of one or more molecules of DNA or RNA, which contain the virus's genes, surrounded by a protein coat called capsid. Some viruses also have an envelope surrounding the capsid. Viruses can be sphere-shaped, or helical. Unlike most bacteria, most viruses do cause disease because they invade living, normal cells, such as those in the human body. They then multiply and produce other viruses like themselves. Each virus is very particular about which cell it attacks.

Various human viruses specifically attack particular cells in the body's organs, systems, or tissues, such as the liver, respiratory system, or blood cells. Although types of viruses behave differently, most survive by taking over the machinery that makes a cell work. Briefly, when a single virus particle, a "virion", comes in contact with a cell it likes, it may attach to special landing sites on the surface of that cell. From there, the virus may inject molecules into the cell, or the cell may swallow up the virion. Once inside the cell, viral molecules such as DNA or RNA direct the cell to make new virus offspring. That's how a virus "infects" a cell. Viruses can even "infect" bacteria. These viruses, called bacteriophages, may help researchers develop alternatives to antibiotic medicines for wiping out bacterial infections. Many viral infections do not result in disease. For example, by the time most people in the United States become adults, they have been infected by cytomegalovirus (CMV). Most of these people, however, do not develop CMV disease symptoms. Other viral infections can result in deadly diseases, such as HIV which causes acquired immunodeficiency syndrome (AIDS) and coronaviruses which cause severe acute respiratory syndrome (SARS).

## Baltimore Classification of Viruses

The Baltimore classification is based on genetic contents and replication strategies of viruses. The genetic material in all types of cells is double-stranded DNA, but some viruses use RNA or single-stranded DNA to carry genetic information.

According to Baltimore classification, viruses are divided into the following seven classes:

1. dsDNA viruses
2. ssDNA viruses

3. dsRNA viruses
4. (+)-sense ssRNA viruses
5. (-)-sense ssRNA viruses
6. RNA reverse transcribing viruses
7. DNA reverse transcribing viruses.

Where "ds" represents "double strand" and "ss" denotes "single strand".

## The Life Cycle of Viruses

The life cycle of viruses may be divided into the following stages:

### *Attachment*

Attachment is a specific binding between viral surface proteins and their receptors on the host cellular surface. This specificity determines the host range of a virus. For instance, the human immunodeficiency virus (HIV) attacks only human's immune cells (mainly T cells), because its surface protein, gp120, can interact with CD4 and chemokine receptors on the T cell's surface.

### *Penetration*

Following attachment, viruses may enter the host cell through receptor mediated endocytosis or other mechanisms.

### *Uncoating*

Uncoating is a process that viral capsid is degraded by viral enzymes or host enzymes.

### *Replication*

Replication involves assembly of viral proteins and genetic materials produced in the host cell.

### *Release*

Viruses may escape from the host cell by causing cell rupture (lysis). Enveloped viruses (e.g., HIV) typically "bud" from the host cell. During the budding process, a virus acquires the phospholipid envelope containing the embedded viral glycoproteins.

# Chapter 5

## The Chemical Basis of Life: Atoms and Molecules

All matter is made up of tiny particles called *atoms*. Atoms are the basic building blocks of all mater, including life. They cannot be broken down by any known means. There are only about 100 kinds of atoms but they can be chemically combined to make thousands of different kinds of substances.

The substances are called *elements* if they are made up of only one kind of atom. Hydrogen gas is an example of an element because there are two atoms of Hydrogen joined together.

Elements are pure sunstances that cannot be broken down by any oridnay means. Unlike solutions, elements are made up of only one kind of atom with the same atomic number. For example, the element Nitrogen consists entirely of Nitrogen atoms. There are over 100 different kinds of elements known today. These elements are organized into different families and different periods in the periodic table.

They can be in the form of any state, solid, liquid, or a gas. Compounds are made up of two or more different *elements* put together in definite proportions. For example, water is a compound because it contains two atoms of Hydrogen for every one atom of Oxygen.

A molecule is like a compound in that it is two or more atoms combined, but each atom when combined has to act as a single particle. If a sodium atom passes an electron to a chlorine atom then the chlorine atom has 17 protons and 18 electrons. An atom that has an excess charge is called an ion.

### *Compounds*

Most substances are compounds. A compound is made up of two or more different kinds of atoms combined in definite proportions. Compounds can be separated into the individual atoms that compose their composition, unlike elements that cannot. Compounds can be seperated only by a chemical change that involves heat or electrcity. For example, Water can be separated int Hydrogen and Oxygen by applying a sufficient DC voltage.

### Ionic Bonding

Ionic bonding is the bond between two atoms formed as a result of the transfer of electrons from one atom to the other. This results in the formation of positive and negative ions. IONS - Any atom which has an excess charge is called an ion, and the force of attraction between two ions is called an ionic bond. Example: When a sodium atom gives up an electron, it will have 11 protons and 10 electrons. This creates a sodium ion which has an excess positive charge of 1 unit. Relate the concepts of solutions, mixtures, acids and bases to the composition and environment of cells.

A solution is a type of mixture that has the same composition throughtout, and appears to be all one substance. In other words, it's a homogeneous mixture that can be in the form of a solid, liquid or gas. The make up of cells consist of many types of mixtures, including solutions. Many of them are in the form of a liquid. A solution is formed by mixing one or more substances into another substance. In any solution, there is a solute and a solvent. The solvent is generally a liquid, and it is the substance in which the solute is dissolved in. A mixture is a word used to describe a blend of different substances. A mixture is made up of two or more substances, in which they can all be visually seperated and distinguished from one another.

### *Solutions*

A solution is any homogeneous mixture, and may include solids, liquids, as well as gases. Most solutions important to living cells are thought of as liquids. The liquid substance that makes up the bulk of a solution is known as a solvent. Substances that dissolve in the solvent are called solutes (SAHL yoots). A solute can either be a solid, liquid, or gas before they are dissolved in the solvent. The most common solutions have water as the solvent.

### Mixtures

A mixture is a combination of substances that are physically mixed without forming new chemical bonds. The substances in a mixture may be present in any proportions. The proportions can change as one substance may be added to, or removed from the mixture, but the different substances in the mixture retain their usual properties.

### Acids

An acid is any compound that produces hydrogen ions in a solution. All acids consist of molecules in which contain hydrogen. The molecules are covalently bonded to another atom or group of atoms. When these molecules dissociate in a solution, such as water, the hydrogen breaks loose as an hydrogen ion. The left over molecules form what is called a negative ion. Hydrochloric acid is an example of an acid, which is important to life activities. This is because the level of acid in the environment of the cell affects the actions of enzymes and therefore affects the functioning of the cell.

### Base

A base is any compound that produces excess hydroxide ions when it is dissolved in a solution. When in a dry state many bases are ionic compounds. A perfect example of a base is sodium hydroxide. Sodium hydroxide is a solid compound, which is made up of sodium and hydroxide ions. When a base is dissolved in water, it dissociates (separates) into its ions:

$$NaOH \longrightarrow Na^+ + OH^-$$

An acid is any compound that produces hydrogen ions in solution. All acids consist of molecules that contain hydrogen covalently bonded to another atom or group of atoms.

A base is a compound that produces hydroxide(hy DRAHK syd) ions (OH-) when disolved in water. In the dry slate, many bases are ionic compounds. Sodium hydroxide (NaOH) is one example. This is a solid compound made of sodium and hydroxide ions. When dissolved in water it separates in to it's ions. NaOH then dissociates to form Na+ and OH- ions.

When acid and a base react their hydrogen and hydroxide combine to form water molecules. This reaction removes these two ions from the solution. that is, they are no longer present as separate ions. However, the negative ions of the acid and the positive ions of the base are still present. Identify the importance of water to life by relating its functions as a transport and dissolving medium.

Water is very important to life. Some scientists beleived that life probaly evolved in water. All living cells are made up of 70-95% water and the water covers 75% of the earths surface. Water is present in all three physical states of mater - solid, liquid, and gas.

Water gets its properties from the fact that it is a polar covalent molecule, and therefore forms hydrogen bonds that connects one water molecule to as many as four other water molecules. This results in a property known as cohesion. Water also has adhesive properties to ionic substances as well as other charged substances such as clay. These properties result in many of water's important properties which include its remarkable ability to dissolve a vast variety of covalent and ionic substances, its high heat capacity, capillarity, surface tension, high heat of vaporization, and the fact that water floats when it freezes. All of these properties of water are related to life on Earth.

Water is considered the universal solvent of life. It acts as a solvent which transports molecules and ions from one area of a cell to another. It also acts as a reactant in chemical reactions that occur in the cell. Without water, molecules and ions would be unable to diffuse through the membrane that surrounds the cell, and gases such as carbon dioxide and oxygen would be unable to pass through plasma membranes.

### *Describe the Characteristics of Organic Compounds*

Organic compounds contain carbon atoms that are linked by covalent bonds to form molecular chains, branches, rings, or sheets. Because there are 4 electrons in the outer energy level of the carbon atom, carbon can combine with other carbon atoms, as well as to the many other biologically important elements such as hydrogen, nitrogen, oxygen, sodium, magnesium, phosphorus, sulfur, chlorine, potassium, calcium, and iron. In addition to carbon, the vast majority of organic compounds also contain hydrogen. Oxides, carbonates and cyanides also contain carbon atoms but are not considered to be organic. Such compounds as carbon monoxide, carbon dioxide, calcium carbonate, and cyanide are considered to be inorganic, even though they contain carbon atoms.

Make an analysis of carbohydrates, lipids, proteins, and nucleic acids in terms of: the atoms that compose these compounds, the function these compounds serve in the body, the food sources of each compound, the sub-units that chemically combine to make these compounds, and examples of each group.

The first compound is carbohydrates. Carbohydrates are sugars and starches. Carbohydrates are found in breads, cereals, potatoes, pastas etc. Lipids is the next compound. Lipids are fats, oils and waxes. Lipids are found in mainly fried foods such as pizza, fries, etc.

Protein is the next compound. Proteins are compounds that contain nitrogen as well as carbon, hydrogen and oxygen. Proteins are found in beans, fish, and meats. The final compound is nucleic acids. Nucleic acids contain phosphorous and nitrogen along with carbon, hydrogen and oxygen. Nucleic acids are mainly found in meats

Carbohydrates are compunds composed of carbon, hydrogen and oxygen. The name carbohydrates comes from the fact that they are formed from the union of carbon dioxide and water during the process of photosynthesis. The ratio of hydrogen to oxygen in a carbohydrate is always two to one - two hydrogen atoms for every oxygen atom. Carbohydrates are normally foun in the breads and cereals as well as fruits and vegetables.

The simplest carbohydrates are just simple sugars (called monosaccharides). These sugars are very important to the body, because they contain large amounts of energy. This energy can be released in the presence of oxygen by breaking down the sugars into carbon dioxide and water.

Lipids include subtances commonly known as fats, waxes and oils. Like carbohydrates, lipids are composed of carbon, hydrogen and oxygen, except there is less oxygen in lipids, than in carbohydrates. Lipids help make up parts of the cell structure (such as the cell membranes) and are used as a reserve for energy in an organism. Lipids provide twice as much energy per gram as the same amount of carbohydrate. Lipids are found in fatty meats such as bacon, as well as products such as whole milk, cream and butter, and from the oils of many plants such as olives, corn, and peanuts.

Proteins are compuounds containing nitrogen, carbon, hydrogen & oxygen. Some proteins also contain sulfur and phosphorus. The number of proteins in your body is virtually countless. Proteins are found everywhere, in cartilage, bones and muscles, hormones, antibodies, and even enzymes. Proteins are notmally found in meats, poltry, fish and some plants such as beans, soy beans, and legumes.

Nucleic Acids are organic compounds that contain Phosphorus, and Nitrogen, as well as Hydrogen, Oxygen and Carbon. Ther are only two known types of nucleic acids, and they are DNA (Deoxyribonucleic Acid) and RNA (Ribonucleic Acid). These two substances were first found in the Nucleus of a Cell. DNA (which is hereditary) and RNA direct and control the activities and structural devolpment of all the cells of an organism.

### *Specify the Importance of Nucleic Acids*

Nucleic Acids are compounds that contain phosphorous and nitrogen in addition to carbon. There are two types of nucleic acids that are formed by the nucleolus located within the nucleus of cells. DNA (deoxyribonucleic acid) and RNA (ribonucleic acid) carry the genetic code that controls the manufacture or synthesis of proteins. These proteins serve as building blocks of cells, as hormones that regulate cellular function, as antibodies that fight infection, and as enzymes that regulate chemical reactions within the cell.

Nucleic acids are a group of organic substances found in the chromosomes of living cells and viruses. They play a central role in the storage and replication of hereditary information and in the expression of this information through protein synthesis. In most organisms, nucleic acids occur in combination with proteins; the combined substances are called nucleoproteins. Nucleic acid molecules are complex chains of varying length. The two chief types of nucleic acids are DNA (deoxyribonucleic acid), which carries the hereditary information from generation to generation, and RNA (ribonucleic acid), which delivers the instructions coded in this information to the cell's protein manufacturing sites (ribosomes).

Outline the function of minerals such as calcium, phosphorus, potassium, and sodium and describe why these substances are essential.

Minerals are chemical elements that organisms need for normal functioning. The most important minerals are iron, calcium, phosphorus, and iodine.

Calcium is the most abundant mineral in the human body, and for that matter, the body of any boned animal, as it is the structural element of bones. Weather our body is deficient in calcium depends less on whether we have a sufficient dietary intake, but rather on how well we absorb it. An important factor in calcium absorption is Vitamin D. The healthy body can produce this vitamin itself for as long as it gets some exposure to sunlight every day. Calcium absorption depends on hormones. Women beyond menopause are under increased risk of developing osteoporosis(bone loss), not because they would lacking in dietary calcium but because they lack the estrogen needed for proper absorption. Trace minerals such as boron also play a part in calcium absorption. The best sources of dietary calcium are milk and milk products, as well as soy and green leafy vegetables. Recommended daily amounts for dietary calcium are about 700 milligram for a healthy adult, and about a gram for adolescents.

Phosphates are essential to the energy-transfer reactions necessary to sustain life processes. Another name for this is adenosine triphospate or ATP which is highly significant to all cells in your body which is involved in every metabolic and photosynthetic reaction. In nature phosphorus occurs only in compounds called phosphates. Plants get phosphorus from soil, and use it in photosynthesis. People and animals take in phosphorus by eating plants and different types of foods.

Distinguish between carbohydrates (sugars and starches), lipids and proteins by using the appropriate tests to identify these nutrients. List each test and how it is conducted.

Tests for lipids: There are two tests that can be used to test for the presence of lipids:

1. Sudan IV : Sudna IV is a special biological stain that is soluble in nonpolar solvents such as lipids and turn from a pink to red colour. Polar compounds will not take up the pink or red colour. For example, if you take a test tube of water and vegetable oil (a lipid) and mix them together and then add a small amount of Sudan IV and shake the mixture, only the oil will turn red or pink not the water.
2. The second test is often called the translucence test for lipid. This is the test that uses unglazed brown paper (same type of brown paper you find in a brown paper bag like the ones from the Co-op used to rap bottles). If lipid is present, the lipid is absorbed into the paper and causes it to appear translucent which means that light is able to pass through and it looks like a greesy spot.

Tests for Protein: The Biuret Test is often used for protein. As you will learn soon, protein is made of a chain of amino acids that are bonded together by what are known as peptide bonds. The Biuret Test uses Biuret Reagent (a special solution) that is able to react with the peptide bonds of the amino acid chain that makes up the protein. This causes a colour change from blue indicating no protein, to pink (small amount of protein), to violet (larger amount of protein), to purple (largest amount of protein). An example to use is to test a known sample of protein such as egg white or gelatin. Adding the Biuret reagent to either of these food materials will cause a purple colour. Adding Biuret Reagent to water would result in blue colour.

Tests for carbohydrate: Benedicts Reagent is used for single sugars (called monosaccharides). Presence of monosaccharides causes the Benedict's solution to change to various colours when heated. The colour change depends on the concentration of the sugar present. Benedict's Reagent is made from Copper Sulfate and therefore contains $Cu^{2+}$ ions. In the presence of simple sugars (known as reducing sugars) the copper is changed to $Cu^{1+}$ ions. The reaction depends on the concentration of the sugar present. Here is a list of colour changes:

- Blue means no sugar present.
- Light green means 0.5% to 1.0% sugar present.

- Green to Yellow means 1.0% to 1.5% sugar present.
- Orange means 1.5% to 2.0% sugar present.
- Red to red brown means greater than 2.0% sugar present.

Iodine is the test for starch. Mixing iodine and starch causes the formation of a starch-iodine complex that is blue-black in colour. If you place iodine on a sample of food and it remains an amber or light brown colour there is no starch present, but if there is starch present, then the colour will change to a dark blue-black colour.

*Explain what vitamins are and describe in general terms the vital body functions of vitamins A, B, C, D, E, and K.*

Vitamins are organic molecules required in the diet in very small amounts. Vitamins are substances that are normally made by plants and then ingested by consumers. The vitamins are important because they serve as coenzymes which help the enzymes function more effectively.

If the body is unable to supply those necessary enzymes in the proper quantities at the proper time, the vitamins and minerals become inert materials and pass unused through your body. Vitamins function directly as coenzymes, or are converted into coenzymes within the cell. Many important biological reactions need vitamins. Vitamin K and some of the B vitamins are essential for blood clotting. This is because Vitmain K is needed for the symthesis of prothrombin, one of the plasma proteins that takes part in the series of reactions leading to the formation of a blood clott.

When large doses of anitbiotics are given to overcome an infection, they can destroy the intestinal bacteria, and a Vitamin K deficiency may result. This is because Vitamin K is produced as a product of the metabolic activities of these bacteria.

### Vitamin A

Vitamin A is a fat-soluble vitamin, which is structurally related to carotene. This is because carotene is converted into vitamin A in the liver. It is composed of compounds called retinoids and carotenoids, and is measured in Retinol Equivalents (RE's) which measure the vitamin A activity of food. There are many vital body functions in which vitamin A is used, some of these are maintaining the protective linings of the lungs, intestine, urinary tract, and other organs, as well as maintaining good vision, and promoting proper bone growth and tooth development. Too little vitamin A leads to a conditon called night blindness. Night blind people have trouble seeing in dim light. Both black-and-white and colour vision make use of the light-sensitive pigment retinal, which is made from Vitamin A. Others include helping in the formation and maintenance of healthy skin and hair, healthy mucus membranes, fortifying the immune system, reproduction, and hormone synthesis and regulation. Also, Vitamin A is used to treat minor skin disorders, it is important for the human metabolism, and is thought to help prevent the development of cancer. Sources of vitamin A include things such as some diary products, eggs, and liver. Some sources of carotene are orange fruits and vegetables, as well as dark, leafy greens. The absence of Vitamin A in ones diet can cause loss in weight, eye diseases (including night blindness), kidney stones, rashes, depression, frequent infections of the respiratory system, bladder, or digestive systems, plus many other problems.

### *Vitamin B1 (Thiamine)*

Vitamin B1 is a water-soluble vitamin which is important for the production of energy and metabolism of sugar. It is also essential for supporting normal appetite, nervous system functions, muscles and heart, as well as aiding in the digestion of carbohydrates, and promoting growth and good muscle tone. Good sources of vitamin B1 are some dairy products, nuts, meats, whole grains, yeast eggs, and germ of cereals. The lack of vitamin B1 in ones diet may cause the loss of appetite, loss of weight, nausea, anxiety, irritability, insomnia, depression, weakness and fatigue. Other problems that may occur are heart and muscle problems, such as enlarged heart, abnormal heart rhythms, heart failure, muscle wasting, aches and pains in muscles, difficulty walking, loss of reflexes, and paralysis.

### *Vitamin B2 (Riboflavin)*

Vitamin B2 is a water-soluble vitamin which is found to release energy from protein, fat, and carbohydrates, and is the precursor of flavoproteins, flavin-adenine dinucleotide and flavin mononucleotide. This vitamin is part of a coenzyme used in energy metabolism, it alleviates eye fatigue, helps maintain cell respiration, healthy eyes, hair, skin, and nails, aids in the formation of antibodies and red blood cells, and promotes general health. Good sources of vitamin B2 are some dairy products, liver, the white of an egg, meats, whole grains, cereals, almonds, and green vegetables. Some problems that are caused by the dietary lack of vitamin B2 are magenta tongue, rashes, cracks and sores in or around the mouth, itching, burning, bloodshot eyes, digestive disturbances, trembling, sluggishness, as well as dry, scaly skin.

### *Vitamin B3 (Niacin)*

Vitamin B3 is a water-soluble vitamin, it's precursor is tryptophan, and it is involved in the oxidative release of energy from food. It is necessary for the metabolism of carbohydrates, is part of a coenzyme used in energy metabolism, helps improve circulation, supports the health of the nervous system, digestive system, and the skin, as well as protects the skin.

Other vital body functions of vitamin B3 are preventing pellagra, and reducing high blood pressure, and help lower cholesterol levels. Dietary sources of vitamin B3 include milk, eggs, meat, whole grains and cereals, nuts, and all protein-containing foods.

Insufficient amounts of vitamin B3 in one's diet may result in pellagra, irritability, nervousness, loss of appetite, dizziness, mental confusion, flaky skin rash on sun exposed areas, and delirium. Others include depression, fatigue, headaches, muscular weakness, insomnia, and black, smooth tongue.

### *Vitamin B5 (Pantothenic Acid)*

Vitamin B5 is a water-soluble vitamin which is essential for the metabolism of food. Some vital body functions of vitamin B5 include metabolism of proteins, carbohydrates, and fats, and the production of steroid hormones and other important chemicals, helps fight depression.

As well as supporting the normal functioning of the gastrointestinal tract, and it is required for the production of cholesterol, bile, red blood cells, and antibodies. Good dietary sources of vitamin B5 are whole grains, beans, liver, eggs, fish, milk and milk products, yeast, lean beef, some potatoes. Others include broccoli, and other vegetables in the cabbage family. Deficiency of vitamin B5 causes problems such as abdominal distress, burning sensation in the heels, and sleep problems.

### *Vitamin B6 (Pyridoxine)*

Vitamin B6 is a water-soluble vitamin, and is related compounds pyridoxamine, and pyridoxal. This vitamin is necessary for the synthesis and breakdown of amino acids, is used for the metabolism of protein, aids in the formation of red blood cells, helps convert tryptophan to niacin, and for maintaining the central nervous system. It also helps maintain a proper balance of sodium and phosphorous in the body, aids in the removal of excess fluid in premenstrual women, promotes healthy skin, and reduces muscle spasms, leg cramps, nausea, as well as hand numbness and stiffness.

Sources of vitamin B6 include bananas, fish, shellfish, meat, poultry, nuts, fruits, whole grains, vegetables, yeast, wheat germ, maize, and rice husks. The chance of deficiency is increased by smoking, and can cause problems such as anaemia, smooth tongue, insomnia, irritability, muscle twitching, convulsions, rashes, kidney stones, greasy dermatitis. In addition, other resulting problems are arm and leg cramps, loss of hair, muscular weakness, slow learning, mouth disorders, water retention, nervousness, and skin eruptions.

### *Vitamin B12 (Cyanocobalamin)*

Vitamin B12 is a water-soluble vitamin, and is involved in the biosynthesis of methyl groups of choline, and methionine. Some vital body functions of vitamin B12 are being part of a coenzyme used in new cell synthesis, help maintain and protect the nerve cells and system, helps in the formation and regeneration of red blood cells, and helps prevent anaemia.

Others include being used for genetic materials, it is necessary for carbohydrate, fat and protein metabolism, promotes growth in children, increases energy, and is needed for calcium absorption. Significant sources of this vitamin are milk, cheese, eggs, liver and meat.

A shortage of vitamin B12 in one's diet can lead to anaemia, poor appetite, growth failure in children, fatigue, depression, lack of balance, smooth tongue, and skin hypersensitivity. In addition, other problems are brain damage, paralysis, nervous system degeneration, as well as degeneration of the spinal cord.

## Define Enzymes and Explain the Make-up and Importance of Enzymes

Enzymes (EN zymz) are protein substances that are necessary for most of the chemical reactions that occur in living cells. Enzymes speed up chemical reactions. Enzymes go into a chemical reaction only temporarily. They are not changed by this reaction so they may be used over and over again for the same chemical step with other molecules. In this way, enzymes are organic catalysts. The substance that an enzyme acts upon is called it's substrate.

Enzymes are almost invisible Protein molecules which are contained in the food we eat and are produced by our bodies. Enzymes are necessary for most of the chemical reations that occur in living cells. All enzymes in an organism are made by the cells of the organism. There are three different types of enzymes which are Metabolic, Digestive, and Food. Enzymes are important because they purify blood, strengthen the immune system, break down fats, lower cholesterol, enhance mental capacity, cleanse the colon, enhance sleep, and improve aging skin.

All living cells make enzymes, but enzymes are not alive. Enzyme molecules function by altering other molecules. Enzymes combine with the altered molecules to form a complex molecule structure in which chemical reactions take place. The enzyme, which remains unchanged, then seprates from the product of the reaction. The human body has thousands of kinds of enzymes. Each kind does it's own job. Without these people could not breath, see, move, or digest food. Photosynthesis in plants also depends on enzymes. There is a major difference between chemical reactions in nonliving things and the reactions in living things. Consider, for example, the burning of gasoline in an automobile engine. The gasoline vapor is admitted to the engine cylinder. It is ignited by a spark. The vapor burns in a fraction of a second. In fact, the burning is so rapid that it produces a small explosion, which helps to drive the engine. How Enzymes work is that the ability to act as a catalysts depends on their shape. Somewhere on the surface of each enzyme there is a region called the active site. An Enzyme may also cause 2 molecules to join. In this case, there are two substrates. Each fits into the active site in such a way that they are brought into close contact. This enables bonds to form between the two substrate molecules. There is a theory to explain how enzyme and substrate fit together at an active site. It is called the lock-and-key model. Just as the notched surface of a key can open only one lock, the shape of the active site of an enzyme only fits certian substrates. Thus, each enzyme can catalyze a reaction only of those substrates.

The more modern view of enzyme function is known as the induced fit model which suggests that the enzyme changes shape just slightly to better fit the substrate.

Enzymes are proteins which act as biological catalysts. A catalyst enables a chemical reaction to take place at lower temperatures by lowering the activation energy required for the reaction to occur.

This is important to living cells since the enzyme enables the metabolic reactions to occur at normal cellular temperatures. Many enzymes must be attached to certain non-protein molecules in order to function. Some molecules are called cofactors, such as copper, iron, and magnesium. Others are called coenzymes.

Many coenzymes consist of vitamins, especially B vitamins. If a persons diet lacks the right amount of these vitamins, the enzymes cannot function properly, and serious problems may occur within the body.

## Molecules of Life: Biomolecules

A biomolecule is any organic molecule that is produced by a living organism, including large polymeric molecules such as proteins, polysaccharides, and nucleic acids as well as small molecules such as primary metabolites, secondary metabolites, and natural products. A more general name for this class of molecules is a biogenic substance.

As organic molecules, biomolecules consist primarily of carbon and hydrogen, nitrogen, and oxygen, and, to a smaller extent, phosphorus and sulfur. Other elements sometimes are incorporated but are much less common.

## Types of Biomolecules

A diverse range of biomolecules exist, including:

- Small molecules:

In the fields of pharmacology and biochemistry, a small molecule is a low molecular weight organic compound which is by definition not a polymer. The term small molecule, especially within the field of pharmacology, is usually restricted to a molecule that also binds with high affinity to a biopolymer such as protein, nucleic acid, or polysaccharide and in addition alters the activity or function of the biopolymer. The upper molecular weight limit for a small molecule is approximately 800 Daltons which allows for the possibility to rapidly diffuse across cell membranes so that they can reach intracellular sites of action. In addition, this molecular weight cutoff is a necessary but insufficient condition for oral bioavailability.

Small molecules can have a variety of biological functions, serving as cell signaling molecules, as tools in molecular biology, as drugs in medicine, as pesticides in farming, and in many other roles. These compounds can be natural (such as secondary metabolites) or artificial (such as antiviral drugs); they may have a beneficial effect against a disease (such as drugs) or may be detrimental (such as teratogens and carcinogens). Biopolymers such as nucleic acids, proteins, and polysaccharides (such as starch or cellulose) are not small molecules, although their constituent monomers — ribo- or deoxyribonucleotides, amino acids, and monosaccharides, respectively — are often considered to be. Very small oligomers are also usually considered small molecules, such as dinucleotides, peptides such as the antioxidant glutathione, and disaccharides such as sucrose.

### *Drugs*

Most drugs are small molecules, although some drugs can be proteins, e.g. insulin. Many proteins are degraded if administered orally and most often cannot cross the cell membranes. Small molecules are more likely to be absorbed, although some of them are only absorbed after oral administration if given as prodrugs. Many dietary supplements are small molecules (but not herb extracts, such as ginkgo).

### *Primary and Secondary Metabolites*

For organisms to produce small molecules they need one or more specialized enzymes (to create and destroy), which as a result are not that varied in vertebrates (recent and small + slow population size), but very common in soil bacteria (such as streptomyces) and fungi, which in particular secrete antibiotics.

Plants also have several secondary metabolites, which play a role in cell signalling, pigmentation or in defence, several of which have also been used as drugs (medical and recreational):

- Alkaloids
- Glycosides
- Lipids
- Flavonoids
- Nonribosomal peptides, such as actinomycin-D
- Phenazines
- Phenols
- Polyketide
- Terpenes, including steroids
- Tetrapyrroles.

### *Research Tools*

Enzymes and receptors are often activated or inhibited by endogenous protein, but can be also inhibited by endogenous or exogenous small molecule inhibitors or activators with can bind to the active site or on the allosteric site.

An example is the teratogen and carcinogen phorbol 12-myristate 13-acetate which is a plant terpene which activates protein kinase C promoting cancer, making it a very useful investigative tool. There is also interest in creating small molecule artificial transcription factors to regulate gene expression, examples include wrenchnol (a wrench shaped molecule). Binding of ligand can be characterised using a variety of analytical techniques such as surface plasmon resonance, microscale thermophoresis or dual polarisation interferometry to quantify the reaction affinities and kinetic properties and also any induced conformational change.

### *Nucleosides and Nucleotides*

Nucleosides are molecules formed by attaching a nucleobase to a ribose ring. Examples of these include cytidine, uridine, adenosine, guanosine, thymidine and inosine.

Nucleosides can be phosphorylated by specific kinases in the cell, producing nucleotides. Both DNA and RNA are polymers, consisting of long, linear molecules. The repeating structural units, or monomers, of the nucleic acids are called nucleotides. Each nucleotide is made of an acyclic nitrogenous base, a pentose and one to three phosphate groups. They contain carbon, nitrogen, oxygen, hydrogen and phosphorus. They serve as sources of chemical energy (adenosine triphosphate and guanosine triphosphate), participate in cellular signaling (cyclic guanosine monophosphate and cyclic adenosine monophosphate), and are incorporated into important cofactors of enzymatic reactions (coenzyme A, flavin adenine dinucleotide, flavin mononucleotide, and nicotinamide adenine dinucleotide phosphate).

### *Saccharides*

Monosaccharides are the simplest form of carbohydrates with only one simple sugar. They essentially contain an aldehyde or ketone group in their structure. The presence of an aldehyde group in a monosaccharide is indicated by the prefix *aldo-*. Similarly, a ketone group is denoted by the prefix *keto-*.

Examples of monosaccharides are the hexoses glucose, fructose, and galactose and pentoses, ribose, and deoxyribose Consumed fructose and glucose have different rates of gastric emptying, are differentially absorbed and have different metabolic fates, providing multiple opportunities for 2 different saccharides to differentially affect food intake. Disaccharides are formed when two monosaccharides, or two single simple sugars, form a bond with removal of water. They can be hydrolyzed to yield their saccharin building blocks by boiling with dilute acid or reacting them with appropriate enzymes. Examples of disaccharides include sucrose, maltose, and lactose. Polysaccharides are polymerized monosaccharides, complex, carbohydrates. They have multiple simple sugars. Examples are starch, cellulose, and glycogen. They are generally large and often have a complex branched connectivity. Because of their size, polysaccharides are not water-soluble, but their many hydroxy groups become hydrated individually when exposed to water, and some polysaccharides form thick colloidal dispersions when heated in water. Shorter polysaccharides, with 3 - 10 monomers, are called oligosaccharides. A fluorescent indicator-displacement molecular imprinting sensor was developed for discriminating saccharides. It successfully discriminated three brands of orange juice beverage. The change in fluorescence intensity of the sensing films resulting is directly related to the saccharide concentration.

### Lignin

Lignin is a random polymer composed mainly of aromatic rings with short (up to three) aliphatic carbons chains connecting the rings. Lignin is the second most common biopolymer (after cellulose) and is one of the primary structural components of most plants. It contains subunits derived from *p*-coumaryl alcohol, coniferyl alcohol, and sinapyl alcohol and is unusual among biomolecules in that it is racemic i.e. it is not optically active. The lack of optical activity is because the polymerization of lignin occurs via free radical coupling reactions in which there is no preference for either configuration at a chiral center.

### Lipids

Lipids are chiefly fatty acid esters, and are the basic building blocks of biological membranes. Another biological role is energy storage (e.g., triglycerides). Most lipids consist of a polar or hydrophilic head (typically glycerol) and one to three nonpolar or hydrophobic fatty acid tails, and therefore they are amphiphilic. Fatty acids consist of unbranched chains of carbon atoms that are connected by single bonds alone (saturated fatty acids) or by both single and double bonds (unsaturated fatty acids). The chains are usually 14-24 carbon groups long, but it is always an even number. For lipids present in biological membranes, the hydrophilic head is from one of three classes:

- Glycolipids, whose heads contain an oligosaccharide with 1-15 saccharide residues.
- Phospholipids, whose heads contain a positively charged group that is linked to the tail by a negatively charged phosphate group.

- Sterols, whose heads contain a planar steroid ring, for example, cholesterol.

Other lipids include prostaglandins and leukotrienes which are both 20-carbon fatty acyl units synthesized from arachidonic acid. They are also known as fatty acids

### Amino Acids

Amino acids contain both amino and carboxylic acid functional groups. (In biochemistry, the term amino acid is used when referring to those amino acids in which the amino and carboxylate functionalities are attached to the same carbon, plus proline which is not actually an amino acid). Amino acids are the building blocks of long polymer chains. With 2-10 amino acids such chains are called peptides, with 10-100 they are often called polypeptides, and longer chains are known as proteins. These protein structures have many structural and enzymatic roles in organisms. There are twenty amino acids that are encoded by the standard genetic code, but there are more than 500 natural amino acids. When amino acids other than the set of twenty are observed in proteins, this is usually the result of modification after translation (protein synthesis). Only two amino acids other than the standard twenty are known to be incorporated into proteins during translation, in certain organisms:

- Selenocysteine is incorporated into some proteins at a UGA codon, which is normally a stop codon.
- Pyrrolysine is incorporated into some proteins at a UAG codon. For instance, in some methanogens in enzymes that are used to produce methane.

Besides those used in protein synthesis, other biologically important amino acids include carnitine (used in lipid transport within a cell), ornithine, GABA and taurine.

### Protein Structure

The particular series of amino acids that form a protein is known as that protein's primary structure. This sequence is determined by the genetic makeup of the individual. Proteins have several, well-classified, elements of local structure formed by intermolecular attraction, this forms the secondary structure of protein.

They are broadly divided in two, alpha helix and beta sheet, also called beta pleated sheets. Alpha helices are formed of coiling of protein due to attraction between amine group of one amino acid with carboxylic acid group of other.

The coil contains about 3.6 amino acids per turn and the alkyl group of amino acid lie outside the plane of coil. Beta pleated sheets are formed by strong continuous hydrogen bond over the length of protein chain. Bonding may be parallel or antiparallel in nature.

Structurally, natural silk is formed of beta pleated sheets. Usually, a protein is formed by action of both these structures in variable ratios. Coiling may also be random. The overall 3D structure of a protein is termed its tertiary structure. It is formed as result of various forces like hydrogen bonding, disulfide bridges, hydrophobic interactions, hydrophilic interactions, van der Waals force etc.

When two or more different polypeptide chains cluster to form a protein, quaternary structure of protein is formed. Quaternary structure is a unique attribute of polymeric and heteromeric proteins like hemoglobin, which consists of two alpha and two beta peptide chains.

### *Apoenzymes*

An apoenzyme is the inactive storage and generally secretory form of a protein. This is required to protect the secretory cell from the activity of that protein.

Apoenzymes becomes active enzyme on addition of a cofactor. Cofactors can be either inorganic (e.g., metal ions and iron-sulfur clusters) or organic compounds, (e.g., flavin and heme). Organic cofactors can be either prosthetic groups, which are tightly bound to an enzyme, or coenzymes, which are released from the enzyme's active site during the reaction.

### *Isoenzymes*

Isoenzymes are enzymes with similar function but different structure. They are products of different genes. They are produced in different organs to perform the same function. LDH are examples of such enzymes. Their varied levels in blood are used to determine any deformity in the organ of secretion.

### *Vitamins*

A vitamin is a compound that is generally not synthesized by a given organism but is nonetheless vital to its survival or health (for example coenzymes). These compounds must be absorbed, or eaten, but typically only in trace quantities. When originally proposed by Casimir Funk, a Polish biochemist, he believed them to all be basic and therefore named them vital amines. The "l" was later dropped to form the word vitamines.

# Chapter 6

# Genetic Material: Properties and Replication

## DNA

Deoxyribonucleic acid (DNA) is a nucleic acid that contains the genetic instructions used in the development and functioning of all known living organisms with the exception of some viruses. The main role of DNA molecules is the long-term storage of information.

DNA is often compared to a set of blueprints, like a recipe or a code, since it contains the instructions needed to construct other components of cells, such as proteins and RNA molecules. The DNA segments that carry this genetic information are called genes, but other DNA sequences have structural purposes, or are involved in regulating the use of this genetic information.

DNA consists of two long polymers of simple units called nucleotides, with backbones made of sugars and phosphate groups joined by ester bonds. These two strands run in opposite directions to each other and are therefore anti-parallel. Attached to each sugar is one of four types of molecules called bases. It is the sequence of these four bases along the backbone that encodes information. This information is read using the genetic code, which specifies the sequence of the amino acids within proteins.

The code is read by copying stretches of DNA into the related nucleic acid RNA, in a process called transcription. Within cells, DNA is organized into long structures called chromosomes. These chromosomes are duplicated before cells divide, in a process called DNA replication. Eukaryotic organisms (animals, plants, fungi, and protists) store most of their DNA inside the cell nucleus and some of their DNA in organelles, such as mitochondria or chloroplasts. In contrast, prokaryotes (bacteria and archaea) store their DNA only in the cytoplasm. Within the chromosomes, chromatin proteins such as histones compact and organize DNA. These compact structures guide the interactions between DNA and other proteins, helping control which parts of the DNA are transcribed.

### Properties

DNA is a long polymer made from repeating units called nucleotides. The DNA chain is 22 to 26 Angströms wide (2.2 to 2.6 nanometres), and one nucleotide unit is 3.3 Å (0.33 nm) long. Although each individual repeating unit is very small, DNA polymers can be very large molecules containing millions of nucleotides. For instance, the largest human chromosome, chromosome number 1, is approximately 220 million base pairs long. In living organisms, DNA does not usually exist as a single molecule, but instead as a pair of molecules that are held tightly together.

These two long strands entwine like vines, in the shape of a double helix. The nucleotide repeats contain both the segment of the backbone of the molecule, which holds the chain together, and a base, which interacts with the other DNA strand in the helix. A base linked to a sugar is called a nucleoside and a base linked to a sugar and one or more phosphate groups is called a nucleotide. If multiple nucleotides are linked together, as in DNA, this polymer is called a polynucleotide.

The backbone of the DNA strand is made from alternating phosphate and sugar residues. The sugar in DNA is 2-deoxyribose, which is a pentose (five-carbon) sugar. The sugars are joined together by phosphate groups that form phosphodiester bonds between the third and fifth carbon atoms of adjacent sugar rings.

These asymmetric bonds mean a strand of DNA has a direction. In a double helix the direction of the nucleotides in one strand is opposite to their direction in the other strand: the strands are *antiparallel*. The asymmetric ends of DNA strands are called the 52 (*five prime*) and 32 (*three prime*) ends, with the 5' end having a terminal phosphate group and the 3' end a terminal hydroxyl group. One major difference between DNA and RNA is the sugar, with the 2-deoxyribose in DNA being replaced by the alternative pentose sugar ribose in RNA.

The DNA double helix is stabilized by hydrogen bonds between the bases attached to the two strands. The four bases found in DNA are adenine (abbreviated A), cytosine (C), guanine (G) and thymine (T). These four bases are attached to the sugar/phosphate to form the complete nucleotide, as shown for adenosine monophosphate.

These bases are classified into two types; adenine and guanine are fused five- and six-membered heterocyclic compounds called purines, while cytosine and thymine are six-membered rings called pyrimidines.

A fifth pyrimidine base, called uracil (U), usually takes the place of thymine in RNA and differs from thymine by lacking a methyl group on its ring. Uracil is not usually found in DNA, occurring only as a breakdown product of cytosine. In addition to RNA and DNA, a large number of artificial nucleic acid analogues have also been created to study the proprieties of nucleic acids, or for use in biotechnology.

### Grooves

Twin helical strands form the DNA backbone. Another double helix may be found by tracing the spaces, or grooves, between the strands. These voids are adjacent to the base pairs and may provide a binding site. As the strands are not directly opposite each other, the grooves are unequally sized. One groove, the major groove, is 22 wide and the other, the minor groove, is 12 wide.

The narrowness of the minor groove means that the edges of the bases are more accessible in the major groove. As a result, proteins like transcription factors that can bind to specific sequences in double-stranded DNA usually make contacts to the sides of the bases exposed in the major groove.

This situation varies in unusual conformations of DNA within the cell, but the major and minor grooves are always named to reflect the differences in size that would be seen if the DNA is twisted back into the ordinary B form.

### *Base Pairing*

Each type of base on one strand forms a bond with just one type of base on the other strand. This is called complementary base pairing. Here, purines form hydrogen bonds to pyrimidines, with A bonding only to T, and C bonding only to G. This arrangement of two nucleotides binding together across the double helix is called a base pair. As hydrogen bonds are not covalent, they can be broken and rejoined relatively easily. The two strands of DNA in a double helix can therefore be pulled apart like a zipper, either by a mechanical force or high temperature. As a result of this complementarity, all the information in the double-stranded sequence of a DNA helix is duplicated on each strand, which is vital in DNA replication. Indeed, this reversible and specific interaction between complementary base pairs is critical for all the functions of DNA in living organisms.

The two types of base pairs form different numbers of hydrogen bonds, AT forming two hydrogen bonds, and GC forming three hydrogen bonds. DNA with high GC-content is more stable than DNA with low GC-content, but contrary to popular belief, this is not due to the extra hydrogen bond of a GC base pair but rather the contribution of stacking interactions (hydrogen bonding merely provides specificity of the pairing, not stability). As a result, it is both the percentage of GC base pairs and the overall length of a DNA double helix that determine the strength of the association between the two strands of DNA. Long DNA helices with a high GC content have stronger-interacting strands, while short helices with high AT content have weaker-interacting strands. In biology, parts of the DNA double helix that need to separate easily, such as the TATAAT Pribnow box in some promoters, tend to have a high AT content, making the strands easier to pull apart. In the laboratory, the strength of this interaction can be measured by finding the temperature required to break the hydrogen bonds, their melting temperature (also called $T_m$ value). When all the base pairs in a DNA double helix melt, the strands separate and exist in solution as two entirely independent molecules. These single-stranded DNA molecules (*ssDNA*) have no single common shape, but some conformations are more stable than others.

### *Sense and Antisense*

A DNA sequence is called "sense" if its sequence is the same as that of a messenger RNA copy that is translated into protein. The sequence on the opposite strand is called the "antisense" sequence. Both sense and antisense sequences can exist on different parts of the same strand of DNA (i.e. both strands contain both sense and antisense sequences). In both prokaryotes and eukaryotes, antisense RNA sequences are produced, but the functions of these RNAs are not entirely clear. One proposal is that antisense RNAs are involved in regulating gene expression through RNA-RNA base pairing.

A few DNA sequences in prokaryotes and eukaryotes, and more in plasmids and viruses, blur the distinction between sense and antisense strands by having overlapping genes. In these cases, some DNA sequences do double duty, encoding one protein when read along one strand, and a second protein when read in the opposite direction along the other strand. In bacteria, this overlap may be involved in the regulation of gene transcription, while in viruses, overlapping genes increase the amount of information that can be encoded within the small viral genome.

### *Supercoiling*

DNA can be twisted like a rope in a process called DNA supercoiling. With DNA in its "relaxed" state, a strand usually circles the axis of the double helix once every 10.4 base pairs, but if the DNA is twisted the strands become more tightly or more loosely wound. If the DNA is twisted in the direction of the helix, this is positive supercoiling, and the bases are held more tightly together. If they are twisted in the opposite direction, this is negative supercoiling, and the bases come apart more easily. In nature, most DNA has slight negative supercoiling that is introduced by enzymes called topoisomerases. These enzymes are also needed to relieve the twisting stresses introduced into DNA strands during processes such as transcription and DNA replication.

### *Alternate DNA Structures*

DNA exists in many possible conformations that include A-DNA, B-DNA, and Z-DNA forms, although, only B-DNA and

Z-DNA have been directly observed in functional organisms. The conformation that DNA adopts depends on the hydration level, DNA sequence, the amount and direction of supercoiling, chemical modifications of the bases, the type and concentration of metal ions, as well as the presence of polyamines in solution.

The first published reports of A-DNA X-ray diffraction patterns— and also B-DNA used analyses based on Patterson transforms that provided only a limited amount of structural information for oriented fibres of DNA. An alternate analysis was then proposed by Wilkins *et al.*, in 1953, for the *in vivo* B-DNA X-ray diffraction/scattering patterns of highly hydrated DNA fibres in terms of squares of Bessel functions. In the same journal, James D. Watson and Francis Crick presented their molecular modeling analysis of the DNA X-ray diffraction patterns to suggest that the structure was a double-helix.

Although the 'B-DNA form' is most common under the conditions found in cells, it is not a well-defined conformation but a family of related DNA conformations that occur at the high hydration levels present in living cells. Their corresponding X-ray diffraction and scattering patterns are characteristic of molecular paracrystals with a significant degree of disorder.

Compared to B-DNA, the A-DNA form is a wider right-handed spiral, with a shallow, wide minor groove and a narrower, deeper major groove. The A form occurs under non-physiological conditions in partially dehydrated samples of DNA, while in the cell it may be produced in hybrid pairings of DNA and RNA strands, as well as in enzyme-DNA complexes. Segments of DNA where the bases have been chemically modified by methylation may undergo a larger change in conformation and adopt the Z form. Here, the strands turn about the helical axis in a left-handed spiral, the opposite of the more common B form. These unusual structures can be recognized by specific Z-DNA binding proteins and may be involved in the regulation of transcription.

### *Quadruplex Structures*

At the ends of the linear chromosomes are specialized regions of DNA called telomeres. The main function of these regions is to allow the cell to replicate chromosome ends using the enzyme telomerase, as the enzymes that normally replicate

DNA cannot copy the extreme 32 ends of chromosomes. These specialized chromosome caps also help protect the DNA ends, and stop the DNA repair systems in the cell from treating them as damage to be corrected. In human cells, telomeres are usually lengths of single-stranded DNA containing several thousand repeats of a simple TTAGGG sequence.

These guanine-rich sequences may stabilize chromosome ends by forming structures of stacked sets of four-base units, rather than the usual base pairs found in other DNA molecules. Here, four guanine bases form a flat plate and these flat four-base units then stack on top of each other, to form a stable *G-quadruplex* structure. These structures are stabilized by hydrogen bonding between the edges of the bases and chelation of a metal ion in the centre of each four-base unit. Other structures can also be formed, with the central set of four bases coming from either a single strand folded around the bases, or several different parallel strands, each contributing one base to the central structure. In addition to these stacked structures, telomeres also form large loop structures called telomere loops, or T-loops. Here, the single-stranded DNA curls around in a long circle stabilized by telomere-binding proteins. At the very end of the T-loop, the single-stranded telomere DNA is held onto a region of double-stranded DNA by the telomere strand disrupting the double-helical DNA and base pairing to one of the two strands. This triple-stranded structure is called a displacement loop or D-loop.

### Branched DNA

In DNA fraying occurs when non-complementary regions exist at the end of an otherwise complementary double-strand of DNA. However, branched DNA can occur if a third strand of DNA is introduced and contains adjoining regions able to hybridize with the frayed regions of the pre-existing double-strand. Although the simplest example of branched DNA involves only three strands of DNA, complexes involving additional strands and multiple branches are also possible.

### Vibration

DNA may carry out low-frequency collective motion as observed by the Raman spectroscopy and analyzed with a quasi-continuum model.

## Chemical Modifications

### Base Modifications

The expression of genes is influenced by how the DNA is packaged in chromosomes, in a structure called chromatin. Base modifications can be involved in packaging, with regions that have low or no gene expression usually containing high levels of methylation of cytosine bases. For example, cytosine methylation, produces 5-methylcytosine, which is important for X-chromosome inactivation. The average level of methylation varies between organisms - the worm *Caenorhabditis elegans* lacks cytosine methylation, while vertebrates have higher levels, with up to 1% of their DNA containing 5-methylcytosine. Despite the importance of 5-methylcytosine, it can deaminate to leave a thymine base, so methylated cytosines are particularly prone to mutations. Other base modifications include adenine methylation in bacteria, the presence of 5-hydroxymethylcytosine in the brain, and the glycosylation of uracil to produce the "J-base" in kinetoplastids.

### Damage

DNA can be damaged by many sorts of mutagens, which change the DNA sequence. Mutagens include oxidizing agents, alkylating agents and also high-energy electromagnetic radiation such as ultraviolet light and X-rays. The type of DNA damage produced depends on the type of mutagen. For example, UV light can damage DNA by producing thymine dimers, which are cross-links between pyrimidine bases.

On the other hand, oxidants such as free radicals or hydrogen peroxide produce multiple forms of damage, including base modifications, particularly of guanosine, and double-strand breaks. A typical human cell contains about 150,000 bases that have suffered oxidative damage. Of these oxidative lesions, the most dangerous are double-strand breaks, as these are difficult to repair and can produce point mutations, insertions and deletions from the DNA sequence, as well as chromosomal translocations.

Many mutagens fit into the space between two adjacent base pairs, this is called *intercalation*. Most intercalators are aromatic and planar molecules; examples include ethidium bromide, daunomycin, and doxorubicin. In order for an intercalator to fit between base pairs, the bases must separate, distorting the DNA strands by unwinding of the double helix. This inhibits both transcription and DNA replication, causing toxicity and mutations.

As a result, DNA intercalators are often carcinogens, and benzo[*a*]pyrene diol epoxide, acridines, aflatoxin and ethidium bromide are well-known examples. Nevertheless, due to their ability to inhibit DNA transcription and replication, other similar toxins are also used in chemotherapy to inhibit rapidly growing cancer cells.

## Biological Functions

DNA usually occurs as linear chromosomes in eukaryotes, and circular chromosomes in prokaryotes. The set of chromosomes in a cell makes up its genome; the human genome has approximately 3 billion base pairs of DNA arranged into 46 chromosomes. The information carried by DNA is held in the sequence of pieces of DNA called genes. Transmission of genetic information in genes is achieved via complementary base pairing.

For example, in transcription, when a cell uses the information in a gene, the DNA sequence is copied into a complementary RNA sequence through the attraction between the DNA and the correct RNA nucleotides. Usually, this RNA copy is then used to make a matching protein sequence in a process called translation, which depends on the same interaction between RNA nucleotides.

In alternative fashion, a cell may simply copy its genetic information in a process called DNA replication. The details of these functions are covered in other articles; here we focus on the interactions between DNA and other molecules that mediate the function of the genome.

### *Genes and Genomes*

Genomic DNA is tightly and orderly packed in the process called DNA condensation to fit the small available volumes of the cell. In eukaryotes, DNA is located in the cell nucleus, as well as small amounts in mitochondria and chloroplasts. In prokaryotes, the DNA is held within an irregularly shaped body in the cytoplasm called the nucleoid. The genetic information in a genome is held within genes, and the complete set of this information in an organism is called its genotype. A gene is a unit of heredity and is a region of DNA that influences a particular characteristic in an organism. Genes contain an open reading frame that can be transcribed, as well as regulatory sequences such as promoters and enhancers, which control the transcription of the open reading frame.

In many species, only a small fraction of the total sequence of the genome encodes protein. For example, only about 1.5% of the human genome consists of protein-coding exons, with over 50% of human DNA consisting of non-coding repetitive sequences. The reasons for the presence of so much non-coding DNA in eukaryotic genomes and the extraordinary differences in genome size, or *C-value*, among species represent a long-standing puzzle known as the "C-value enigma". However, DNA sequences that do not code protein may still encode functional non-coding RNA molecules, which are involved in the regulation of gene expression.

Some non-coding DNA sequences play structural roles in chromosomes. Telomeres and centromeres typically contain few genes, but are important for the function and stability of chromosomes. An abundant form of non-coding DNA in humans are pseudogenes, which are copies of genes that have been disabled by mutation. These sequences are usually just molecular fossils, although they can occasionally serve as raw genetic material for the creation of new genes through the process of gene duplication and divergence.

### *Transcription and Translation*

A gene is a sequence of DNA that contains genetic information and can influence the phenotype of an organism. Within a gene, the sequence of bases along a DNA strand defines a messenger RNA sequence, which then defines one or more protein sequences.

The relationship between the nucleotide sequences of genes and the amino-acid sequences of proteins is determined by the rules of translation, known collectively as the genetic code. The genetic code consists of three-letter 'words' called *codons* formed from a sequence of three nucleotides (e.g. ACT, CAG, TTT). In transcription, the codons of a gene are copied into messenger

RNA by RNA polymerase. This RNA copy is then decoded by a ribosome that reads the RNA sequence by base-pairing the messenger RNA to transfer RNA, which carries amino acids. Since there are 4 bases in 3-letter combinations, there are 64 possible codons (4 combinations). These encode the twenty standard amino acids, giving most amino acids more than one possible codon. There are also three 'stop' or 'nonsense' codons signifying the end of the coding region; these are the TAA, TGA and TAG codons.

### *Replication*

Cell division is essential for an organism to grow, but, when a cell divides, it must replicate the DNA in its genome so that the two daughter cells have the same genetic information as their parent. The double-stranded structure of DNA provides a simple mechanism for DNA replication. Here, the two strands are separated and then each strand's complementary DNA sequence is recreated by an enzyme called DNA polymerase. This enzyme makes the complementary strand by finding the correct base through complementary base pairing, and bonding it onto the original strand. As DNA polymerases can only extend a DNA strand in a 52 to 32 direction, different mechanisms are used to copy the antiparallel strands of the double helix. In this way, the base on the old strand dictates which base appears on the new strand, and the cell ends up with a perfect copy of its DNA.

## Interactions with Proteins

All the functions of DNA depend on interactions with proteins. These protein interactions can be non-specific, or the protein can bind specifically to a single DNA sequence. Enzymes can also bind to DNA and of these, the polymerases that copy the DNA base sequence in transcription and DNA replication are particularly important.

### *DNA-binding Proteins*

Structural proteins that bind DNA are well-understood examples of non-specific DNA-protein interactions. Within chromosomes, DNA is held in complexes with structural proteins. These proteins organize the DNA into a compact structure called chromatin. In eukaryotes this structure involves

DNA binding to a complex of small basic proteins called histones, while in prokaryotes multiple types of proteins are involved. The histones form a disk-shaped complex called a nucleosome, which contains two complete turns of double-stranded DNA wrapped around its surface.

These non-specific interactions are formed through basic residues in the histones making ionic bonds to the acidic sugar-phosphate backbone of the DNA, and are therefore largely independent of the base sequence. Chemical modifications of these basic amino acid residues include methylation, phosphorylation and acetylation. These chemical changes alter the strength of the interaction between the DNA and the histones, making the DNA more or less accessible to transcription factors and changing the rate of transcription. Other non-specific DNA-binding proteins in chromatin include the high-mobility group proteins, which bind to bent or distorted DNA. These proteins are important in bending arrays of nucleosomes and arranging them into the larger structures that make up chromosomes.

A distinct group of DNA-binding proteins are the DNA-binding proteins that specifically bind single-stranded DNA. In humans, replication protein A is the best-understood member of this family and is used in processes where the double helix is separated, including DNA replication, recombination and DNA repair. These binding proteins seem to stabilize single-stranded DNA and protect it from forming stem-loops or being degraded by nucleases. In contrast, other proteins have evolved to bind to particular DNA sequences. The most intensively studied of these are the various transcription factors, which are proteins that regulate transcription. Each transcription factor binds to one particular set of DNA sequences and activates or inhibits the transcription of genes that have these sequences close to their promoters. The transcription factors do this in two ways. Firstly, they can bind the RNA polymerase responsible for transcription, either directly or through other mediator proteins; this locates the polymerase at the promoter and allows it to begin transcription. Alternatively, transcription factors can bind enzymes that modify the histones at the promoter; this will change the accessibility of the DNA template to the polymerase.

As these DNA targets can occur throughout an organism's genome, changes in the activity of one type of transcription factor can affect thousands of genes. Consequently, these proteins are often the targets of the signal transduction processes that control responses to environmental changes or cellular differentiation and development. The specificity of these transcription factors' interactions with DNA come from the proteins making multiple contacts to the edges of the DNA bases, allowing them to "read" the DNA sequence. Most of these base-interactions are made in the major groove, where the bases are most accessible.

## DNA-modifying Enzymes

### *Nucleases and Ligases*

Nucleases are enzymes that cut DNA strands by catalyzing the hydrolysis of the phosphodiester bonds. Nucleases that hydrolyse nucleotides from the ends of DNA strands are called exonucleases, while endonucleases cut within strands. The most frequently used nucleases in molecular biology are the restriction endonucleases, which cut DNA at specific sequences. For instance, the EcoRV enzyme shown to the left recognizes the 6-base sequence 52 -GAT|ATC-32 and makes a cut at the vertical line. In nature, these enzymes protect bacteria against phage infection by digesting the phage DNA when it enters the bacterial cell, acting as part of the restriction modification system. In technology, these sequence-specific nucleases are used in molecular cloning and DNA fingerprinting.

Enzymes called DNA ligases can rejoin cut or broken DNA strands. Ligases are particularly important in lagging strand DNA replication, as they join together the short segments of DNA produced at the replication fork into a complete copy of the DNA template. They are also used in DNA repair and genetic recombination.

### *Topoisomerases and Helicases*

Topoisomerases are enzymes with both nuclease and ligase activity. These proteins change the amount of supercoiling in DNA. Some of these enzymes work by cutting the DNA helix and allowing one section to rotate, thereby reducing its level of supercoiling; the enzyme then seals the DNA break. Other types of these enzymes are capable of cutting one DNA helix and then passing a second strand of DNA through this break, before rejoining the helix. Topoisomerases are required for many processes involving DNA, such as DNA replication and transcription. Helicases are proteins that are a type of molecular motor. They use the chemical energy in nucleoside triphosphates, predominantly ATP, to break hydrogen bonds between bases and unwind the DNA double helix into single strands. These enzymes are essential for most processes where enzymes need to access the DNA bases.

### *Polymerases*

Polymerases are enzymes that synthesize polynucleotide chains from nucleoside triphosphates. The sequence of their products are copies of existing polynucleotide chains - which are called *templates*. These enzymes function by adding nucleotides onto the 32 hydroxyl group of the previous nucleotide in a DNA strand. As a consequence, all polymerases work in a 52 to 32 direction. In the active site of these enzymes, the incoming nucleoside triphosphate base-pairs to the template: this allows polymerases to accurately synthesize the complementary strand of their template. Polymerases are classified according to the type of template that they use.

In DNA replication, a DNA-dependent DNA polymerase makes a copy of a DNA sequence. Accuracy is vital in this process, so many of these polymerases have a proofreading activity. Here, the polymerase recognizes the occasional mistakes in the synthesis reaction by the lack of base pairing between the mismatched nucleotides. If a mismatch is detected, a 32 to 52 exonuclease activity is activated and the incorrect base removed. In most organisms, DNA polymerases function in a large complex called the replisome that contains multiple accessory subunits, such as the DNA clamp or helicases.

RNA-dependent DNA polymerases are a specialized class of polymerases that copy the sequence of an RNA strand into DNA. They include reverse transcriptase, which is a viral enzyme involved in the infection of cells by retroviruses, and telomerase, which is required for the replication of telomeres. Telomerase is an unusual polymerase because it contains its own RNA template as part of its structure.

Transcription is carried out by a DNA-dependent RNA polymerase that copies the sequence of a DNA strand into RNA. To begin transcribing a gene, the RNA polymerase binds to a sequence of DNA called a promoter and separates the DNA strands. It then copies the gene sequence into a messenger RNA transcript until it reaches a region of DNA called the terminator, where it halts and detaches from the DNA. As with human DNA-dependent DNA polymerases, RNA polymerase II, the enzyme that transcribes most of the genes in the human genome, operates as part of a large protein complex with multiple regulatory and accessory subunits.

### *Genetic Recombination*

A DNA helix usually does not interact with other segments of DNA, and in human cells the different chromosomes even occupy separate areas in the nucleus called "chromosome territories". This physical separation of different chromosomes is important for the ability of DNA to function as a stable repository for information, as one of the few times chromosomes interact is during chromosomal crossover when they recombine. Chromosomal crossover is when two DNA helices break, swap a section and then rejoin.

Recombination allows chromosomes to exchange genetic information and produces new combinations of genes, which increases the efficiency of natural selection and can be important in the rapid evolution of new proteins. Genetic recombination can also be involved in DNA repair, particularly in the cell's response to double-strand breaks.

The most common form of chromosomal crossover is homologous recombination, where the two chromosomes involved share very similar sequences. Non-homologous recombination can be damaging to cells, as it can produce chromosomal translocations and genetic abnormalities. The recombination reaction is catalyzed by enzymes known as recombinases, such as RAD51.

The first step in recombination is a double-stranded break either caused by an endonuclease or damage to the DNA. A series of steps catalyzed in part by the recombinase then leads to joining of the two helices by at least one Holliday junction, in which a segment of a single strand in each helix is annealed to the complementary strand in the other helix. The Holliday junction is a tetrahedral junction structure that can be moved along the pair of chromosomes, swapping one strand for another. The recombination reaction is then halted by cleavage of the junction and re-ligation of the released DNA.

### *Evolution*

DNA contains the genetic information that allows all modern living things to function, grow and reproduce. However, it is unclear how long in the 4-billion-year history of life DNA has performed this function, as it has been proposed that the earliest forms of life may have used RNA as their genetic material. RNA may have acted as the central part of early cell metabolism as it can both transmit genetic information and carry out catalysis as part of ribozymes. This ancient RNA world where nucleic acid would have been used for both catalysis and genetics may have influenced the evolution of the current genetic code based on four nucleotide bases. This would occur, since the number of different bases in such an organism is a trade-off between a small number of bases increasing replication accuracy and a large number of bases increasing the catalytic efficiency of ribozymes.

However, there is no direct evidence of ancient genetic systems, as recovery of DNA from most fossils is impossible. This is because DNA will survive in the environment for less than one million years and slowly degrades into short fragments in solution. Claims for older DNA have been made, most notably a report of the isolation of a viable bacterium from a salt crystal 250 million years old, but these claims are controversial.

## Uses in Technology

### *Genetic Engineering*

Methods have been developed to purify DNA from organisms, such as phenol-chloroform extraction and manipulate it in the laboratory, such as restriction digests and the polymerase chain reaction. Modern biology and biochemistry make intensive use of these techniques in recombinant DNA technology. Recombinant DNA is a man-made DNA sequence that has been assembled from other DNA sequences. They can be transformed into organisms in the form of plasmids or in the appropriate format, by using a viral vector. The genetically modified organisms produced can be used to produce products such as recombinant proteins, used in medical research, or be grown in agriculture.

### *Forensics*

Forensic scientists can use DNA in blood, semen, skin, saliva or hair found at a crime scene to identify a matching DNA of an individual, such as a perpetrator. This process is formally termed DNA profiling, but may also be called "genetic fingerprinting". In DNA profiling, the lengths of variable sections of repetitive DNA, such as short tandem repeats and minisatellites, are compared between people.

This method is usually an extremely reliable technique for identifying a matching DNA. However, identification can be complicated if the scene is contaminated with DNA from several people. DNA profiling was developed in 1984 by British geneticist Sir Alec Jeffreys, and first used in forensic science to convict Colin Pitchfork in the 1988 Enderby murders case.

People convicted of certain types of crimes may be required to provide a sample of DNA for a database. This has helped investigators solve old cases where only a DNA sample was obtained from the scene. DNA profiling can also be used to identify victims of mass casualty incidents. On the other hand, many convicted people have been released from prison on the basis of DNA techniques, which were not available when a crime had originally been committed.

## DNA Profiling

DNA profiling (also called DNA testing, DNA typing, or genetic fingerprinting) is a technique employed by forensic scientists to assist in the identification of individuals on the basis of their respective DNA profiles. DNA profiles are encrypted sets of numbers that reflect a person's DNA make-up, which can also be used as the person's identifier. DNA profiling should not be confused with full genome sequencing. It is used in, for example, parental testing and rape investigation.

Although 99.9% of human DNA sequences are the same in every person, enough of the DNA is different to distinguish one individual from another. DNA profiling uses repetitive ("repeat") sequences that are highly variable, called variable number tandem repeats (VNTR). VNTRs loci are very similar between closely related humans, but so variable that unrelated individuals are extremely unlikely to have the same VNTRs. The DNA profiling technique was first reported in 1984 by Sir Alec Jeffreys at the University of Leicester in England, and is now the basis of several national DNA databases. Dr. Jeffreys's genetic fingerprinting was made commercially available in 1987, when a chemical company, ICI, started a blood-testing centre in England.

### *DNA Profiling Process*

The process begins with a sample of an individual's DNA (typically called a "reference sample"). The most desirable method of collecting a reference sample is the use of a buccal swab, as this reduces the possibility of contamination.

When this is not available (e.g. because a court order may be needed and not obtainable) other methods may need to be used to collect a sample of blood, saliva, semen, or other appropriate fluid or tissue from personal items (e.g. toothbrush, razor, etc.) or from stored samples (e.g. banked sperm or biopsy tissue). Samples obtained from blood relatives (biological relative) can provide an indication of an individual's profile, as could human remains which had been previously profiled.

A reference sample is then analyzed to create the individual's DNA profile using one of a number of techniques, discussed below. The DNA profile is then compared against another sample to determine whether there is a genetic match.

### *RFLP Analysis*

The first methods for finding out genetics used for DNA profiling involved restriction enzyme digestion, followed by Southern blot analysis. Although polymorphisms can exist in the restriction enzyme cleavage sites, more commonly the enzymes and DNA probes were used to analyze VNTR loci. However, the Southern blot technique is laborious, and requires large amounts of undegraded sample DNA.

Also, Karl Brown's original technique looked at many minisatellite loci at the same time, increasing the observed variability, but making it hard to discern individual alleles (and thereby precluding parental testing). These early techniques have been supplanted by PCR-based assays.

### *PCR Analysis*

With the invention of the polymerase chain reaction (PCR) technique, DNA profiling took huge strides forward in both discriminating power and the ability to recover information from very small (or degraded) starting samples. PCR greatly amplifies the amounts of a specific region of DNA, using oligonucleotide primers and a thermostable DNA polymerase.

Early assays such as the HLA-DQ alpha reverse dot blot strips grew to be very popular due to their ease of use, and the speed with which a result could be obtained. However they were not as discriminating as RFLP. It was also difficult to determine a DNA profile for mixed samples, such as a vaginal swab from a sexual assault victim.

Fortunately, the PCR method is readily adaptable for analyzing VNTR loci. In the United States the FBI has standardized a set of 13 VNTR assays for DNA typing, and has organized the CODIS database for forensic identification in criminal cases. Similar assays and databases have been set up in other countries. Also, commercial kits are available that analyze single-nucleotide polymorphisms (SNPs). These kits use PCR to amplify specific regions with known variations and hybridize them to probes anchored on cards, which results in a coloured spot corresponding to the particular sequence variation.

### *STR Analysis*

The method of DNA profiling used today is based on PCR and uses short tandem repeats (STR). This method uses highly polymorphic regions that have short repeated sequences of DNA (the most common is 4 bases repeated, but there are other lengths in use, including 3 and 5 bases).

Because unrelated people almost certainly have different numbers of repeat units, STRs can be used to discriminate between unrelated individuals. These STR loci (locations on a chromosome) are targeted with sequence-specific primers and amplified using PCR. The DNA fragments that result are then separated and detected using electrophoresis. There are two common methods of separation and detection, capillary electrophoresis (CE) and gel electrophoresis.

Each STR is polymorphic, however, the number of alleles is small. Typically each STR allele will be shared by around 5 - 20% of individuals. The power of STR analysis comes from looking at multiple STR loci simultaneously. The pattern of alleles can identify an individual quite accurately. Thus STR analysis provides an excellent identification tool. The more STR regions that are tested in an individual the more discriminating the test becomes.

From country to country, different STR-based DNA-profiling systems are in use. In North America, systems which amplify the CODIS 13 core loci are almost universal, while in the UK the SGM+ system (which is compatible with The National DNA Database), is in use. Whichever system is used, many of the STR regions used are the same. These DNA-profiling systems are based on multiplex reactions, whereby many STR regions will be tested at the same time. The true power of STR analysis is in its statistical power of discrimination. Because the 13 loci that are currently used for discrimination in CODIS are independently assorted (having a certain number of repeats at one locus doesn't change the likelihood of having any number of repeats at any other locus), the product rule for probabilities can be applied. This means that if someone has the DNA type of ABC, where the three loci were independent, we can say that the probability of having that DNA type is the probability of having type A times the probability of having type B times the probability of having type C. This has resulted in the ability to generate match probabilities of 1 in a quintillion (1 with 18 zeros after it) or more. However, DNA database searches showed much more frequent than expected false DNA matches including one perfect 13 locus match out of only 30,000 DNA samples in Maryland in January 2007.

Moreover, since there are about 12 million monozygotic twins on Earth, that theoretical probability is useless. For example, the actual probability that 2 random people have the same DNA depends on whether there were twins or triplets (etc.) in the family, and the number of loci used in the test. Where twins are common, the probability of matching the DNA is 22 in 1000, or about 2.2 in 100 will have matching DNA.

In practice, the risk of contaminated-matching is much greater than matching a distant relative, such as a sample being contaminated from nearby objects, or from left-over cells transferred from a prior test. Logically, the risk is greater for matching the most common person in the samples: everything collected from, or in contact with, a victim is a major source of contamination for any other samples brought into a lab. For that reason, multiple control-samples are typically tested, to ensure that they stayed clean, when prepared during the same period as the actual test samples. Unexpected matches (or variations) in several control-samples indicates a high probability of contamination for the actual test samples. In a relationship test, the full DNA profiles should differ (except for twins), to prove that a person wasn't actually matched as being related to their own DNA in another sample.

### AmpFLP

Another technique, AmpFLP, or amplified fragment length polymorphism was also put into practice during the early 1990s. This technique was also faster than RFLP analysis and used PCR to amplify DNA samples. It relied on variable number tandem repeat (VNTR) polymorphisms to distinguish various alleles, which were separated on a polyacrylamide gel using an allelic ladder (as opposed to a molecular weight ladder). Bands could be visualized by silver staining the gel. One popular locus for fingerprinting was the D1S80 locus.

As with all PCR based methods, highly degraded DNA or very small amounts of DNA may cause allelic dropout (causing a mistake in thinking a heterozygote is a homozygote) or other stochastic effects. In addition, because the analysis is done on a gel, very high number repeats may bunch together at the top of the gel, making it difficult to resolve. AmpFLP analysis can be highly automated, and allows for easy creation of phylogenetic trees based on comparing individual samples of DNA. Due to its relatively low cost and ease of set-up and operation, AmpFLP remains popular in lower income countries.

### DNA Family Relationship Analysis

Using PCR technology, DNA analysis is widely applied to determine genetic family relationships such as paternity, maternity, siblingship and other kinships.

During conception, the father's sperm cell and the mother's egg cell, each containing half the amount of DNA found in other body cells, meet and fuse to form a fertilized egg, called a zygote. The zygote contains a complete set of DNA molecules, a unique combination of DNA from both parents. This zygote divides and multiplies into an embryo and later, a full human being. At each stage of development, all the cells forming the body contain the same DNA—half from the father and half from the mother. This fact allows the relationship testing to use all types of all samples including loose cells from the cheeks collected using buccal swabs, blood or other types of samples.

While a lot of DNA contains information for a certain function, there is some called junk DNA, which is currently used for human identification. At some special locations (called loci) in the junk DNA, predictable inheritance patterns were found to be useful in determining biological relationships. These locations contain specific DNA markers that DNA scientists use to identify individuals. In a routine DNA paternity test, the markers used are Short Tandem Repeats (STRs), short pieces of DNA that occur in highly differential repeat patterns among individuals. Each person's DNA contains two copies of these markers—one copy inherited from the father and one from the mother. Within a population, the markers at each person's DNA location could differ in length and sometimes sequence, depending on the markers inherited from the parents. The combination of marker sizes found in each person makes up his/her unique genetic profile. When determining the relationship between two individuals, their genetic profiles are compared to see if they share the same inheritance patterns at a statistically conclusive rate. For example, the following sample report from this commercial DNA paternity testing laboratory Universal Genetics signifies how relatedness between parents and child is identified on those special markers:

| *DNA Marker* | *Mother* | *Child* | *Alleged father* |
|---|---|---|---|
| D21S11 | 28, 30 | 28, 31 | 29, 31 |
| D7S820 | 9, 10 | 10, 11 | 11, 12 |
| TH01 | 14, 15 | 14, 16 | 15, 16 |
| D13S317 | 7, 8 | 7, 9 | 8, 9 |
| D19S433 | 14, 16.2 | 14, 15 | 15, 17 |

The partial results indicate that the child and the alleged father's DNA match among these five markers. The complete test results show this correlation on 16 markers between the child and the tested man to draw a conclusion of whether or not the man is the biological father.

Scientifically, each marker is assigned with a Paternity Index (PI), which is a statistical measure of how powerfully a match at a particular marker indicates paternity.

The PI of each marker is multiplied with each other to generate the Combined Paternity Index (CPI), which indicates the overall probability of an individual being the biological father of the tested child relative to any random man from the entire population of the same race. The CPI is then converted into a Probability of Paternity showing the degree of relatedness between the alleged father and child.

The DNA test report in other family relationship tests, such as grandparentage and siblingship tests, is similar to a paternity test report. Instead of the Combined Paternity Index, a different value, such as a Siblingship Index, is reported.

The report shows the genetic profiles of each tested person. If there are markers shared among the tested individuals, the probability of biological relationship is calculated to determine how likely the tested individuals share the same markers due to a blood relationship.

### *Y-chromosome Analysis*

Recent innovations have included the creation of primers targeting polymorphic regions on the Y-chromosome (Y-STR), which allows resolution of a mixed DNA sample from a male and female and/or cases in which a differential extraction is not possible. Y-chromosomes are paternally inherited, so Y-STR analysis can help in the identification of paternally related males. Y-STR analysis was performed in the Sally Hemings controversy to determine if Thomas Jefferson had sired a son with one of his slaves.

### *Mitochondrial Analysis*

For highly degraded samples, it is sometimes impossible to get a complete profile of the 13 CODIS STRs. In these situations, mitochondrial DNA (mtDNA) is sometimes typed due to there being many copies of mtDNA in a cell, while there may only be 1-2 copies of the nuclear DNA. Forensic scientists amplify the HV1 and HV2 regions of the mtDNA, then sequence each region and compare single-nucleotide differences to a reference. Because mtDNA is maternally inherited, directly linked maternal relatives can be used as match references, such as one's maternal grandmother's daughter's son.

A difference of two or more nucleotides is generally considered to be an exclusion. Heteroplasmy and poly-C differences may throw off straight sequence comparisons, so some expertise on the part of the analyst is required. mtDNA is useful in determining clear identities, such as those of missing people when a maternally linked relative can be found. mtDNA testing was used in determining that Anna Anderson was not the Russian princess she had claimed to be, Anastasia Romanov.

mtDNA can be obtained from such material as hair shafts and old bones/teeth..

### *DNA Databases*

There are now several DNA databases in existence around the world. Some are private, but most of the largest databases are government controlled. The United States maintains the largest DNA database, with the Combined DNA Index System, holding over 5 million records as of 2007.

The United Kingdom maintains the National DNA Database (NDNAD), which is of similar size, despite the UK's smaller population. The size of this database, and its rate of growth, is giving concern to civil liberties groups in the UK, where police have wide-ranging powers to take samples and retain them even in the event of acquittal.

The U.S. Patriot Act of the United States provides a means for the U.S. government to get DNA samples from other countries if they are either a division of, or head office of, a company operating in the U.S.

Under the act, the American offices of the company can't divulge to their subsidiaries/offices in other countries the reasons that these DNA samples are sought or by whom.

When a match is made from a National DNA Databank to link a crime scene to an offender who has provided a DNA

Sample to a databank that link is often referred to as a *cold hit.* A cold hit is of value in referring the police agency to a specific suspect but is of less evidential value than a DNA match made from outside the DNA Databank.

### *Considerations when Evaluating DNA Evidence*

In the early days of the use of genetic fingerprinting as criminal evidence, juries were often swayed by spurious statistical arguments by defence lawyers along these lines: given a match that had a 1 in 5 million probability of occurring by chance, the lawyer would argue that this meant that in a country of say 60 million people there were 12 people who would also match the profile. This was then translated to a 1 in 12 chance of the suspect being the guilty one.

This argument is not sound unless the suspect was drawn at random from the population of the country. In fact, a jury should consider how likely it is that an individual matching the genetic profile would also have been a suspect in the case for other reasons. Another spurious statistical argument is based on the false assumption that a 1 in 5 million probability of a match automatically translates into a 1 in 5 million probability of innocence and is known as the prosecutor's fallacy.

When using RFLP, the theoretical risk of a coincidental match is 1 in 100 billion (100,000,000,000), although the practical risk is actually 1 in 1000 because monozygotic twins are 0.2% of the human population. Moreover, the rate of laboratory error is almost certainly higher than this, and often actual laboratory procedures do not reflect the theory under which the coincidence probabilities were computed.

For example, the coincidence probabilities may be calculated based on the probabilities that markers in two samples have bands in *precisely* the same location, but a laboratory worker may conclude that similar—but not precisely identical—band patterns result from identical genetic samples with some imperfection in the agarose gel. However, in this case, the laboratory worker increases the coincidence risk by expanding the criteria for declaring a match. Recent studies have quoted relatively high error rates which may be cause for concern. In the early days of genetic fingerprinting, the necessary population data to accurately compute a match probability was sometimes unavailable. Between 1992 and 1996, arbitrary low ceilings were controversially put on match probabilities used in RFLP analysis rather than the higher theoretically computed ones. Today, RFLP has become widely disused due to the advent of more discriminating, sensitive and easier technologies. STRs do not suffer from such subjectivity and provide similar power of discrimination (1 in 10^13 for unrelated individuals if using a full SGM+ profile) It should be noted that figures of this magnitude are not considered to be statistically supportable by scientists in the UK, for unrelated individuals with full matching DNA profiles a match probability of 1 in a billion is considered statistically supportable (Since 1998 the DNA profiling system supported by The National DNA Database in the UK is the SGM+ DNA profiling system which includes 10 STR regions and a sex indicating test. However, with any DNA technique, the cautious juror should not convict on genetic fingerprint evidence alone if other factors raise doubt. Contamination with other evidence (secondary transfer) is a key source of incorrect DNA profiles and raising doubts as to whether a sample has been adulterated is a favorite defence technique. More rarely, chimerism is one such instance where the lack of a genetic match may unfairly exclude a suspect.

### Evidence of Genetic Relationship

It's also possible to use DNA profiling as evidence of genetic relationship, but testing that shows no relationship isn't absolutely certain. While almost all individuals have a single and distinct set of genes, rare individuals, known as "chimeras", have at least two different sets of genes. There have been several cases of DNA profiling that falsely "proved" that a mother was unrelated to her children.

### Fake DNA Evidence

The value of DNA evidence has to be seen in light of recent cases where criminals planted fake DNA samples at crime scenes. In one case, a criminal even planted fake DNA evidence in his own body: Dr. John Schneeberger raped one of his sedated patients in 1992 and left semen on her underwear. Police drew what they believed to be Schneeberger's blood and compared its DNA against the crime scene semen DNA on three occasions, never showing a match. It turned out that he had surgically inserted a Penrose drain into his arm and filled it with foreign blood and anticoagulants.

In a study conducted by the life science company Nucleix and published in the journal Forensic Science International, scientists found that an In vitro synthesized sample of DNA matching any desired genetic profile can be constructed using standard molecular biology techniques without obtaining any actual tissue from that person.

## DNA Evidence as Evidence in Criminal Trials

### *Familial Searching*

Familial searching is the use of family members' DNA to identify a closely related suspect in jurisdictions where large DNA databases exist, but no exact match has been found. The first successful use of the practice was in a UK case where a man was convicted of manslaughter when he threw a brick stained with his own blood into a moving car. Police could not get an exact match to the UK's DNA database because the man had no criminal convictions, but police implicated him using a close relative's DNA. The technique was used to catch a Los Angeles serial killer known as the "Grim Sleeper" in 2010. However, critics have challenged the technology as "deeply antithetical to American values" and likely to lead to excess investigation of racial and ethnic minorities.

### *Surreptitious DNA Collecting*

Police forces may collect DNA samples without the suspects' knowledge, and use it as evidence. Legality of this mode of proceeding has been questioned in Australia. In the United States, it has been accepted, courts often claiming that there was no expectation of privacy, citing California v. Greenwood (1985), during which the Supreme Court held that the Fourth Amendment does not prohibit the warrantless search and seizure of garbage left for collection outside the curtilage of a home. Critics of this practice underline the fact that this analogy ignores that "most people have no idea that they risk surrendering their genetic identity to the police by, for instance, failing to destroy a used coffee cup. Moreover, even if they do realize it, there is no way to avoid abandoning one's DNA in public." In the UK, the Human Tissue Act 2004 prohibited private individuals from covertly collecting biological samples (hair, fingernails, etc.) for DNA analysis, but excluded medical and criminal investigations from the offense.

### England and Wales

Evidence from an expert who has compared DNA samples must be accompanied by evidence as to the sources of the samples and the procedures for obtaining the DNA profiles. The judge must ensure that the jury must understand the significance of DNA matches and mismatches in the profiles. The judge must also ensure that the jury does not confuse the 'match probability' (the probability that a person that is chosen at random has a matching DNA profile to the sample from the scene) with the 'likelihood ratio' (the probability that a person with matching DNA committed the crime).

In *R v. Doheny*, EWCA Crim 728. Phillips LJ gave this example of a summing up, which should be carefully tailored to the particular facts in each case:

Members of the Jury, if you accept the scientific evidence called by the Crown, this indicates that there are probably only four or five white males in the United Kingdom from whom that semen stain could have come. The Defendant is one of them. If that is the position, the decision you have to reach, on all the evidence, is whether you are sure that it was the Defendant who left that stain or whether it is possible that it was one of that other small group of men who share the same DNA characteristics. Juries should weigh up conflicting and corroborative evidence, using their own common sense and not by using mathematical formulae, such as Bayes' theorem, so as to avoid "confusion, misunderstanding and misjudgment".

### Presentation and Evaluation of Evidence of Partial or Incomplete DNA Profiles

R v Bates (2006) EWCA Crim 1395 Moore-Bick LJ said: "We can see no reason why partial profile DNA evidence should not be admissible provided that the jury are made aware of its inherent limitations and are given a sufficient explanation to enable them to evaluate it.

There may be cases where the match probability in relation to all the samples tested is so great that the judge would consider its probative value to be minimal and decide to exclude the evidence in the exercise of his discretion, but this gives rise to no new question of principle and can be left for decision on a case by case basis.

However, the fact that there exists in the case of all partial profile evidence the possibility that a "missing" allele might exculpate the accused altogether does not provide sufficient grounds for rejecting such evidence.

In many there is a possibility (at least in theory) that evidence exists which would assist the accused and perhaps even exculpate him altogether, but that does not provide grounds for excluding relevant evidence that is available and otherwise admissible, though it does make it important to ensure that the jury are given sufficient information to enable them to evaluate that evidence properly".

### DNA Testing in the US

There are state laws on DNA profiling in all 50 states of the United States. Detailed information on database laws in each state can be found at the National Conference of State Legislatures website.

### Development of Artificial DNA

In August 2009, scientists in Israel stunned the forensic sciences and raised serious questions concerning the use of DNA by law enforcement as the ultimate method of identification. In a paper published in the journal *Forensic Science International: Genetics*, the Israeli researchers demonstrated that it is possible to manufacture DNA in a laboratory, and thus falsify DNA evidence. The scientists fabricated saliva and blood samples, which originally contained DNA from a person other than the ostensible donor of the blood and saliva.

The researchers also showed that, using a DNA database, it is possible to take information from a profile and manufacture DNA to match it, and that this can be done without access to any actual DNA from the person whose DNA they are duplicating. The synthetic DNA oligos required for the procedure are common in molecular laboratories. Dr. Daniel Frumkin, lead author on the paper, was quoted in *The New York Times* as saying, "You can just engineer a crime scene... any biology undergraduate could perform this." Dr. Frumkin perfected a test that can forensically differentiate real DNA samples from fake ones. His test detects epigenetic modifications, in particular, DNA methylation. Seventy percent of the DNA in any human genome is methylated, meaning it contains methyl group modifications within a CpG dinucleotide context. Methylation at the promoter region is associated with gene silencing. The synthetic DNA lacks this epigenetic modification, which allows the test to distinguish manufactured DNA from original, genuine, DNA.

It is unknown how many, if any, police departments currently use the test, which appears to be a serious issue in light of Frumkin's claim that the DNA manufacturing procedure is within the grasp of any undergraduate biology student. No police lab has publicly announced that it is using the new test to verify DNA results, while FSI Genetics says that any forensic laboratory doing DNA identification should adopt this test to authenticate its results as "real" DNA.

## Bioinformatics

Bioinformatics involves the manipulation, searching, and data mining of DNA sequence data. The development of techniques to store and search DNA sequences have led to widely applied advances in computer science, especially string searching algorithms, machine learning and database theory. String searching or matching algorithms, which find an occurrence of a sequence of letters inside a larger sequence of letters, were developed to search for specific sequences of nucleotides. In other applications such as text editors, even simple algorithms for this problem usually suffice, but DNA sequences cause these algorithms to exhibit near-worst-case behaviour due to their small number of distinct characters.

The related problem of sequence alignment aims to identify homologous sequences and locate the specific mutations that make them distinct. These techniques, especially multiple sequence alignment, are used in studying phylogenetic relationships and protein function. Data sets representing entire genomes' worth of DNA sequences, such as those produced by the Human Genome Project, are difficult to use without annotations, which label the locations of genes and regulatory elements on each chromosome. Regions of DNA sequence that have the characteristic patterns associated with protein- or RNA-coding genes can be identified by gene finding algorithms, which allow researchers to predict the presence of particular gene products in an organism even before they have been isolated experimentally.

### DNA Nanotechnology

DNA nanotechnology uses the unique molecular recognition properties of DNA and other nucleic acids to create self-assembling branched DNA complexes with useful properties. DNA is thus used as a structural material rather than as a carrier of biological information.

This has led to the creation of two-dimensional periodic lattices (both tile-based as well as using the "DNA origami" method) as well as three-dimensional structures in the shapes of polyhedra. Nanomechanical devices and algorithmic self-assembly have also been demonstrated, and these DNA structures have been used to template the arrangement of other molecules such as gold nanoparticles and streptavidin proteins.

### History and Anthropology

Because DNA collects mutations over time, which are then inherited, it contains historical information, and, by comparing DNA sequences, geneticists can infer the evolutionary history of organisms, their phylogeny. This field of phylogenetics is a powerful tool in evolutionary biology. If DNA sequences within a species are compared, population geneticists can learn the history of particular populations. This can be used in studies ranging from ecological genetics to anthropology; For example, DNA evidence is being used to try to identify the Ten Lost Tribes of Israel.

DNA has also been used to look at modern family relationships, such as establishing family relationships between the descendants of Sally Hemings and Thomas Jefferson. This usage is closely related to the use of DNA in criminal investigations detailed above. Indeed, some criminal investigations have been solved when DNA from crime scenes has matched relatives of the guilty individual.

## History of DNA Research

DNA was first isolated by the Swiss physician Friedrich Miescher who, in 1869, discovered a microscopic substance in the pus of discarded surgical bandages. As it resided in the nuclei of cells, he called it "nuclein". In 1919, Phoebus Levene identified the base, sugar and phosphate nucleotide unit. Levene suggested that DNA consisted of a string of nucleotide units linked together through the phosphate groups. However, Levene thought the chain was short and the bases repeated in a fixed order. In 1937 William Astbury produced the first X-ray diffraction patterns that showed that DNA had a regular structure. In 1928, Frederick Griffith discovered that traits of the "smooth" form of the *Pneumococcus* could be transferred to the "rough" form of the same bacteria by mixing killed "smooth" bacteria with the live "rough" form. This system provided the first clear suggestion that DNA carries genetic information—the Avery-MacLeod-McCarty experiment—when Oswald Avery, along with coworkers Colin MacLeod and Maclyn McCarty, identified DNA as the transforming principle in 1943. DNA's role in heredity was confirmed in 1952, when Alfred Hershey and Martha Chase in the Hershey-Chase experiment showed that DNA is the genetic material of the T2 phage.

In 1953, James D. Watson and Francis Crick suggested what is now accepted as the first correct double-helix model of DNA structure in the journal *Nature*. Their double-helix, molecular model of DNA was then based on a single X-ray diffraction image (labeled as "Photo 51") taken by Rosalind Franklin and Raymond Gosling in May 1952, as well as the information that the DNA bases are paired — also obtained through private communications from Erwin Chargaff in the previous years. Chargaff's rules played a very important role in establishing double-helix configurations for B-DNA as well as A-DNA. Experimental evidence supporting the Watson and Crick model were published in a series of five articles in the same issue of *Nature*. Of these, Franklin and Gosling's paper was the first publication of their own X-ray diffraction data and original analysis method that partially supported the Watson and Crick model; this issue also contained an article on DNA structure by Maurice Wilkins and two of his colleagues, whose analysis and *in vivo* B-DNA X-ray patterns also supported the presence *in vivo* of the double-helical DNA configurations as proposed by Crick and Watson for their double-helix molecular model of DNA in the previous two pages of *Nature*. In 1962, after Franklin's death, Watson, Crick, and Wilkins jointly received the Nobel Prize in Physiology or Medicine. However, Nobel rules of the time allowed only living recipients, but a vigorous debate continues on who should receive credit for the discovery.

In an influential presentation in 1957, Crick laid out the "Central Dogma" of molecular biology, which foretold the relationship between DNA, RNA, and proteins, and articulated the "adaptor hypothesis". Final confirmation of the replication mechanism that was implied by the double-helical structure followed in 1958 through the Meselson-Stahl experiment. Further work by Crick and coworkers showed that the genetic code was based on non-overlapping triplets of bases, called codons, allowing Har Gobind Khorana, Robert W. Holley and Marshall Warren Nirenberg to decipher the genetic code. These findings represent the birth of molecular biology.

## Protein Synthesis

Protein synthesis is the process in which cells build proteins. The term is sometimes used to refer only to protein translation but more often it refers to a multi-step process, beginning with amino acid synthesis and transcription of nuclear DNA into messenger RNA, which is then used as input to translation.

The cistron DNA is transcribed into a variety of RNA intermediates. The last version is used as a template in synthesis of a polypeptide chain. Proteins can often be synthesized directly from genes by translating mRNA. When a protein needs to be available on short notice or in large quantities, a protein precursor is produced. A proprotein is an inactive protein containing one or more inhibitory peptides that can be activated when the inhibitory sequence is removed by proteolysis during posttranslational modification. A preprotein is a form that contains a signal sequence (an N-terminal signal peptide) that specifies its insertion into or through membranes; i.e., targets them for secretion. The signal peptide is cleaved off in the endoplasmic reticulum. Preproproteins have both sequences (inhibitory and signal) still present.

For synthesis of protein, a succession of tRNA molecules charged with appropriate amino acids have to be brought together with an mRNA molecule and matched up by base-pairing through their anti-codons with each of its successive codons. The amino acids then have to be linked together to extend the growing protein chain, and the tRNAs, relieved of their burdens, have to be released. This whole complex of processes is carried out by a giant multimolecular machine, the ribosome, formed of two main chains of RNA, called ribosomal RNA (rRNA), and more than 50 different proteins.

This molecular juggernaut latches onto the end of an mRNA molecule and then trundles along it, capturing loaded tRNA molecules and stitching together the amino acids they carry to form a new protein chain. Protein biosynthesis, although very similar, is different for prokaryotes and eukaryotes.

## Amino Acid Synthesis

Amino acid synthesis is the set of biochemical processes (metabolic pathways) by which the various amino acids are produced from other compounds. The substrates for these processes are various compounds in the organism's diet or growth media. Not all organisms are able to synthesise all amino acids, for example humans are only able to synthesise 12 of the 20 standard amino acids.

A fundamental problem for biological systems is to obtain nitrogen in an easily usable form. This problem is solved by certain microorganisms capable of reducing the inert Na"N molecule (nitrogen gas) to two molecules of ammonia in one of the most remarkable reactions in biochemistry.

Nitrogen in the form of ammonia is the source of nitrogen for all the amino acids. The carbon backbones come from the glycolytic pathway, the pentose phosphate pathway, or the citric acid cycle.

In amino acid production, one encounters an important problem in biosynthesis— namely, stereochemical control. Because all amino acids except glycine are chiral, biosynthetic pathways must generate the correct isomer with high fidelity.

In each of the 19 pathways for the generation of chiral amino acids, the stereochemistry at the á-carbon atom is established by a transamination reaction that involves pyridoxal phosphate. Almost all the transaminases that catalyze these reactions descend from a common ancestor, illustrating once again that effective solutions to biochemical problems are retained throughout evolution.

Biosynthetic pathways are often highly regulated such that building blocks are synthesized only when supplies are low. Very often, a high concentration of the final product of a pathway inhibits the activity of enzymes that function early in the pathway. Often present are allosteric enzymes capable of sensing and responding to concentrations of regulatory species. These enzymes are similar in functional properties to aspartate transcarbamoylase and its regulators. Feedback and allosteric mechanisms ensure that all twenty amino acids are maintained in sufficient amounts for protein synthesis and other processes.

Amino acids are synthesized from α-ketoacids, and later transaminated from another aminoacid, usually Glutamate. The enzyme involved in this reaction is an aminotransferase.

Glutamate itself is formed by amination of α-ketoglutarate:

### Nitrogen Fixation: Microorganisms Use ATP and a Powerful Reductant to Reduce Atmospheric Nitrogen to Ammonia

Microorganisms use ATP and reduced ferredoxin, a powerful reductant, to reduce N2 to NH3. An iron-molybdenum cluster in nitrogenase deftly catalyzes the fixation of N2, a very inert molecule. Higher organisms consume the fixed nitrogen to synthesize amino acids, nucleotides, and other nitrogen-containing biomolecules. The major points of entry of NH4+ into metabolism are glutamine or glutamate.

### Amino Acids are Made from Intermediates of the Citric Acid Cycle and Other Major Pathways

Of the basic set of 20 amino acids (not counting selenocysteine) there are 8 that human beings cannot synthesize. In addition, the amino acids arginine, cysteine, glycine, glutamine, histidine, proline, serine and tyrosine are considered conditionally essential, meaning they are not normally required in the diet, but must be supplied exogenously to specific populations that do not synthesize it in adequate amounts. (For example, enough arginine is synthesized by the urea cycle to meet the needs of an adult but perhaps not those of a growing child.) Amino acids that need to be obtained from the diet are called essential amino acids. Nonessential amino acids are produced in the body. The pathways for the synthesis of nonessential amino acids are quite simple. Glutamate dehydrogenase catalyzes the reductive amination of α-ketoglutarate to glutamate. A transamination reaction takes place in the synthesis of most amino acids. At this step, the chirality of the amino acid is established. Alanine and aspartate are synthesized by the transamination of pyruvate and oxaloacetate, respectively. Glutamine is synthesized from NH4+ and glutamate, and asparagine is synthesized similarly. Proline and arginine are derived from glutamate.

Serine, formed from 3-phosphoglycerate, is the precursor of glycine and cysteine. Tyrosine is synthesized by the hydroxylation of phenylalanine, an essential amino acid. The pathways for the biosynthesis of essential amino acids are much more complex than those for the nonessential ones.

Tetrahydrofolate, a carrier of activated one-carbon units, plays an important role in the metabolism of amino acids and nucleotides. This coenzyme carries one-carbon units at three oxidation states, which are interconvertible: most reduced—methyl; intermediate—methylene; and most oxidized—formyl, formimino, and methenyl. The major donor of activated methyl groups is S-adenosylmethionine, which is synthesized by the transfer of an adenosyl group from ATP to the sulfur atom of methionine. S-Adenosylhomocysteine is formed when the activated methyl group is transferred to an acceptor. It is hydrolyzed to adenosine and homocysteine, the latter of which is then methylated to methionine to complete the activated methyl cycle. The A Cortisol inhibites protein synthesis.

### *Amino Acid Biosynthesis is Regulated by Feedback Inhibition*

Most of the pathways of amino acid biosynthesis are regulated by feedback inhibition, in which the committed step is allosterically inhibited by the final product. Branched pathways require extensive interaction among the branches that includes both negative and positive regulation. The regulation of glutamine synthetase from E. coli is a striking demonstration of cumulative feedback inhibition and of control by a cascade of reversible covalent modifications.

### *Amino Acids are Precursors of Many Biomolecules*

Amino acids are precursors of a variety of biomolecules. Glutathione serves as a sulfhydryl buffer and detoxifying agent. Glutathione peroxidase, a selenoenzyme, catalyzes the reduction of hydrogen peroxide and organic peroxides by glutathione. Nitric oxide, a short-lived messenger, is formed from arginine. Porphyrins are synthesized from glycine and succinyl CoA, which condense to give β-aminolevulinate. Two molecules of this intermediate become linked to form porphobilinogen. Four molecules of porphobilinogen combine to form a linear tetrapyrrole, which cyclizes to uroporphyrinogen III. Oxidation and side-chain modifications lead to the synthesis of protoporphyrin IX, which acquires an iron atom to form heme.

## Transcription

In transcription an mRNA chain is generated, with one strand of the DNA double helix in the genome as template. This strand is called the template strand. Transcription can be divided into 3 stages: Initiation, Elongation and Termination, each regulated by a large number of proteins such as transcription factors and coactivators that ensure the correct gene is transcribed.

The DNA strand is read in the 3' to 5' direction and the mRNA is transcribed in the 5' to 3' direction by the RNA polymerase.

Transcription occurs in the cell nucleus, where the DNA is held. The DNA structure of the cell is made up of two helixes made up of sugar and phosphate held together by the bases. The sugar and the phosphate are joined together by covalent bond. The DNA is "unzipped" by the enzyme helicase, leaving the single nucleotide chain open to be copied. RNA polymerase reads the DNA strand from 3 prime (3') end to the 5 prime (5') end, while it synthesizes a single strand of messenger RNA in the 5' to 3' direction. The general RNA structure is very similar to the DNA structure, but in RNA the nucleotide uracil takes the place that thymine occupies in DNA. The single strand of mRNA leaves the nucleus through nuclear pores, and migrates into the cytoplasm. The first product of transcription differs in prokaryotic cells from that of eukaryotic cells, as in prokaryotic cells the product is mRNA, which needs no post-transcriptional modification, while in eukaryotic cells, the first product is called primary transcript, that needs post-transcriptional modification (capping with 7 methyl guanosine, tailing with a poly A tail) to give hnRNA (heterophil nuclear RNA). hnRNA then undergoes splicing of introns (non coding parts of the gene) via spliceosomes to produce the final mRNA.

### *Translation*

The synthesis of proteins is known as translation. Translation occurs in the cytoplasm, where the ribosomes are located. Ribosomes are made of a small and large subunit that surround the mRNA. In translation, messenger RNA (mRNA) is decoded to produce a specific polypeptide according to the rules specified by the trinucleotide genetic code.

This uses an mRNA sequence as a template to guide the synthesis of a chain of amino acids that form a protein. Translation proceeds in four phases: activation, initiation, elongation, and termination (all describing the growth of the amino acid chain, or polypeptide that is the product of translation).

In activation, the correct amino acid (AA) is joined to the correct transfer RNA (tRNA). While this is not technically a step in translation, it is required for translation to proceed. The AA is joined by its carboxyl group to the 3' OH of the tRNA by an ester bond. When the tRNA has an amino acid linked to it, it is termed "charged". Initiation involves the small subunit of the ribosome binding to 5' end of mRNA with the help of initiation factors (IF), other proteins that assist the process. Elongation occurs when the next aminoacyl-tRNA (charged tRNA) in line binds to the ribosome along with GTP and an elongation factor.

Termination of the polypeptide happens when the A site of the ribosome faces a stop codon (UAA, UAG, or UGA). When this happens, no tRNA can recognize it, but releasing factor can recognize nonsense codons and causes the release of the polypeptide chain. The capacity of disabling or inhibiting translation in protein biosynthesis is used by antibiotics such as: anisomycin, cycloheximide, chloramphenicol, tetracycline, streptomycin, erythromycin, puromycin etc.

Translation is the process of converting the mRNA codon sequences into an amino acid polypeptide chain.

1. Amino acid activation
2. Initiation - A ribosome attaches to the mRNA and starts to code at the FMet codon (usually AUG, sometimes GUG or UUG).
3. Elongation - tRNA brings the corresponding amino acid (which has an anticodon that identifies the amino acid as the corresponding molecule to a codon) to each codon as the ribosome moves down the mRNA strand.
4. Termination - Reading of the final mRNA codon (aka the STOP codon), which ends the synthesis of the peptide chain and releases it.

### *Events Following Protein Translation*

The events following biosynthesis include post-translational modification and protein folding. During and after synthesis, polypeptide chains often fold to assume, so called, native secondary and tertiary structures. This is known as *protein folding*.

Many proteins undergo *post-translational modification*. This may include the formation of disulfide bridges or attachment of any of a number of biochemical functional groups, such as acetate, phosphate, various lipids and carbohydrates. Enzymes may also remove one or more amino acids from the leading (amino) end of the polypeptide chain, leaving a protein consisting of two polypeptide chains connected by disulfide bonds.

In general, protein molecules are believed to be modified by small chemical groups, post-translationally. Chemical modifications such as, phosphorylation of serine/threonine, acetylation or methylation of lysine, hydroxylation of proline/ lysine, formylation of glycine, glycosylation of serine/threonine/ asparagine, acylation of cysteine, myristoylation of glycine, biotinylation of lysine, ubiquitination, etc. on proteins is a very important issue in relation to properly understanding the biological functions of a given protein. These much-studied post-translational modifications have become well-established with the discovery of the respective enzymes (kinases for phosphorylation, acetylases for acetylation, methyl transferases for methylation, etc.), which carry on the chemical modifications on the specific amino acid residues. All of these modifications are still believed to have happened as post-translational events. There is no study yet on when actually one particular modification occurs on a given amino acid residue in a given protein. Does it happen when the protein is already formed, or when the amino acid chain is being synthesized, or before the translation of the primary chain has begun?

Since these chemical modifications are related to the biological functions of a protein, it is easy to think that these chemical modifications have happened to the whole protein molecule, after the protein primary chain is fully synthesized; but, if that is the case, we have to consider the fact that the primary chains get folded instantly, (in a similar way as the newly synthesized DNA strands form helixes), to attain its compact-globular conformation ; As most of the primary chains are fairly long [a 5Kd protein may have 40-45 amino acid residues in its primary chain], it is likely that the newly formed amino acid chain tries to remain intact by folding, thereby avoiding its breakdown via lots of proteases present within the cytoplasm. And, no capping event to protect the N-terminal end of the primary sequence (similar to 5' m-RNA capping to protect m-RNAs) is ever discovered for protein primary structure. So, by folding mechanism, the primary chain, perhaps, avoids the protease attacks.

However, once it gets folded, it may be very difficult for the respective enzyme molecule to find out the particular aa residue from the complexity of that compactly folded conformation. In addition, it can be clearly imagined that this enzymatic modification/reaction on a given amino acid requires presence and association of the appropriate enzyme, necessary cofactors, etc. This association is much easier to occur when the amino acid residues in the primary structure are readily available for binding; in other words, it is much difficult for the enzyme molecules to find and to bind to its substrate amino acid residue in a mature protein molecule after its three-dimensional conformation been attained.

So one can think that the modifications can happen while the primary chain of the protein is being synthesized during the translation process on the m-RNA strand; the amino acid residues on the primary chain can be modified instantly and enzymatically by kinases, acetylases, hydroxylases, methyltransferases, etc. to initiate proper folding for protection in order to avoid degradation by proteases, thereby gaining the globular form.

Also, the reader can imagine another scenario in which, while the free amino acid molecules are formed within a cell and become available, they [in that free state] may be modified enzymatically before taking the ride to the translational event; this means that, while the primary chain is being synthesized, the pre-modified amino acid molecules are ready to be engaged. This way, as the primary chain is getting synthesized, it does not have to be modified by the modifying enzymes anymore, and can fold itself instantaneously without concern of degradation. This also arises new thoughts that (i) phosphate, acetyl, methyl, biotinyl, acyl, etc. groups may have the ability to inhibit protease actions on the primary chains; (ii) most or all of the amino acid residues get modified by small chemical groups (so far, only some chemical groups are known).

It is still not fully known exactly when and how the actual modification of a given amino acid residue occurs, at which stage of synthesis, within a protein molecule. Once the chemically-modified, protease-insensitive, intact protein molecule is generated, it must perform its biological function, which requires its being activated. The activation is probably done by a second set of enzymatic reactions when these chemical groups are removed from the a residues (or added back). So, the second type of post-translational modifications are the opposite reactions of the above-described type, which are dephosphorylation by phosphatases or phosphoryl transferases, deacetylation by deacetylases or acetyltransferases, demethylation by demethylases or methyl-transferases, ubiquitination, SuMoylation, glycosylation, biotinylation, etc. These are taking place on the whole protein molecule toward generating its activated form to execute a particular function or toward its deactivation followed by its total degradation.

# Chapter 7

# Genetic Disorders

A genetic disorder is an illness caused by abnormalities in genes or chromosomes. While some diseases, such as cancer, are due in part to genetic disorders, they can also be caused by environmental factors. Most disorders are quite rare and affect one person in every several thousands or millions. Some types of recessive gene disorders confer an advantage in the heterozygous state in certain environments.

## Single Gene Disorder

A single gene disorder is the result of a single mutated gene. There are estimated to be over 4000 human diseases caused by single gene defects. Single gene disorders can be passed on to subsequent generations in several ways. Genomic imprinting and uniparental disomy, however, may affect inheritance patterns. The divisions between recessive and dominant types are not "hard and fast" although the divisions between autosomal and X-linked types are (since the latter types are distinguished purely based on the chromosomal location of the gene).

For example, achondroplasia is typically considered a dominant disorder, but children with two genes for achondroplasia have a severe skeletal disorder that achondroplasics could be viewed as carriers of. Sickle-cell anemia is also considered a recessive condition, but heterozygous carriers have increased resistance to malaria in early childhood, which could be described as a related dominant condition. When a couple where one partner or both are sufferers or carriers of a single gene disorder and wish to have a child they can do so through IVF whichs means they can then have PGD (pre-implantation genetic diagnosis) to check whether the fertilised egg has had the genetic disorder passed on. Autosomal Dominant. Only one mutated copy of the gene will be necessary for a person to be affected by an autosomal dominant disorder. Each affected person usually has one affected parent. There is a 50% chance that a child will inherit the mutated gene. Conditions that are autosomal dominant often have low penetrance, which means that although only one mutated copy is needed, a relatively small proportion of those who inherit that mutation go on to develop the disease. Examples of this type of disorder are Huntington's disease, neurofibromatosis type 1, Marfan syndrome, hereditary nonpolyposis colorectal cancer, and hereditary multiple exostoses, which is a highly penetrant autosomal dominant disorder. Birth defects are also called congenital anomalies.

### *Autosomal Recessive*

Two copies of the gene must be mutated for a person to be affected by an autosomal recessive disorder. An affected person usually has unaffected parents who each carry a single copy of the mutated gene (and are referred to as carriers). Two unaffected people who each carry one copy of the mutated gene have a 25% chance with each pregnancy of having a child affected by the disorder. Examples of this type of disorder are cystic fibrosis, sickle-cell disease (also partial sickle-cell disease), Tay-Sachs disease, Niemann-Pick disease, spinal muscular atrophy, Roberts syndrome, and Dry (otherwise known as "rice-brand") earwax.

### *X-linked Dominant*

X-linked dominant disorders are caused by mutations in genes on the X chromosome. Only a few disorders have this inheritance pattern, with a prime example being X-linked hypophosphatemic rickets. Males and females are both affected in these disorders, with males typically being more severely affected than females. Some X-linked dominant conditions such as Rett syndrome, incontinentia pigmenti type 2 and Aicardi syndrome are usually fatal in males either in utero or shortly after birth, and are therefore predominantly seen in females. Exceptions to this finding are extremely rare cases in which boys with Klinefelter syndrome (47,XXY) also inherit an X-linked dominant condition and exhibit symptoms more similar to those of a female in terms of disease severity. The chance of passing on an X-linked dominant disorder differs between men and women. The sons of a man with an X-linked dominant disorder will all be unaffected (since they receive their father's Y chromosome), and his daughters will all inherit the condition. A woman with an X-linked dominant disorder has a 50% chance of having an affected fetus with each pregnancy, although it should be noted that in cases such as incontinentia pigmenti only female offspring are generally viable. In addition, although these conditions do not alter fertility per se, individuals with Rett syndrome or Aicardi syndrome rarely reproduce.

### *X-linked Recessive*

X-linked recessive conditions are also caused by mutations in genes on the X chromosome. Males are more frequently affected than females, and the chance of passing on the disorder differs between men and women. The sons of a man with an X-linked recessive disorder will not be affected, and his daughters will carry one copy of the mutated gene. A woman who is a carrier of an X-linked recessive disorder (XX) has a 50% chance of having sons who are affected and a 50% chance of having daughters who carry one copy of the mutated gene and are therefore carriers. X-linked recessive conditions include the serious diseases Hemophilia A, Duchenne muscular dystrophy, and Lesch-Nyhan syndrome as well as common and less serious conditions such as male pattern baldness and red-green color blindness. X-linked recessive conditions can sometimes manifest in females due to skewed X-inactivation or monosomy X (Turner syndrome).

### *Y-linked*

Y-linked disorders are caused by mutations on the Y chromosome. Because males inherit a Y chromosome from their fathers, *every* son of an affected father will be affected. Because females inherit an X chromosome from their fathers, female offspring of affected fathers are *never* affected. Since the Y chromosome is relatively small and contains very few genes, there are relatively few Y-linked disorders. Often the symptoms include infertility, which may be circumvented with the help of some fertility treatments. Examples are male infertility and hypertrichosis pinnae.

### *Mitochondrial*

This type of inheritance, also known as maternal inheritance, applies to genes in mitochondrial DNA. Because only egg cells contribute mitochondria to the developing embryo, only mothers can pass on mitochondrial conditions to their children. An example of this type of disorder is Leber's hereditary optic neuropathy. underwear

## Multifactorial and Polygenic (Complex) Disorders

Genetic disorders may also be complex, multifactorial, or polygenic, meaning that they are likely associated with the effects of multiple genes in combination with lifestyle and environmental factors. Multifactorial disorders include heart disease and diabetes. Although complex disorders often cluster in families, they do not have a clear-cut pattern of inheritance. This makes it difficult to determine a person's risk of inheriting or passing on these disorders. Complex disorders are also difficult to study and treat because the specific factors that cause most of these disorders have not yet been identified.

On a pedigree, polygenic diseases do tend to "run in families", but the inheritance does not fit simple patterns as with Mendelian diseases. But this does not mean that the genes cannot eventually be located and studied. There is also a strong environmental component to many of them (e.g., blood pressure).

- asthma
- autoimmune diseases such as multiple sclerosis
- cancers
- ciliopathies
- cleft palate
- diabetes
- heart disease
- hypertension
- inflammatory bowel disease
- mental retardation
- mood disorder

- obesity
- refractive error.

### *Prognosis and Treatment of Genetic Disorders*

Genetic disorders rarely have effective treatments, though gene therapy is being tested as a possible treatment for some genetic diseases, including some forms of retinitis pigmentosa

- Gauchers disease is a genetic disease affecting metabolism. It is more treatable then most other genetic diseases, and can be treated with enzyme replacement therapy, medication miglustat, and bone marrow transplantion.

## Hypothyroidism

### *What is Hypothyroidism?*

The thyroid is the largest endocrine gland in the body. It sits just below the larynx (voice box) and wraps around the trachea (windpipe). The thyroid gland produces thyroid hormone, which helps the body grow and develop. It also plays an important role in the body's metabolism (the processes in the body that use energy, such as eating, breathing, and regulating heat). Hypothyroidism (or underactive thyroid) is a common condition in which the thyroid gland produces too little thyroid hormone. About 1 in 5,000 babies is born with congenital hypothyroidism, in which the thyroid fails to grow normally and cannot produce enough of its hormone. There is no known cause for most cases of congenital hypothyroidism. But about 10 to 20 percent of the time, the condition is caused by an inherited defect that alters the production of thyroid hormone. The most common inherited form of hypothyroidism is a defect of the TPO (thyroid peroxidase) gene on chromosome 2. This gene plays an important role in thyroid hormone production.

### *How do People get Hypothyroidism?*

Hypothyroidism may be caused by 1) an autoimmune disease that attacks the thyroid gland, 2) surgery or radiation to treat thyroid cancer and other conditions, or 3) rare and random genetic events in which a mutation is acquired during early development.

### *What are the Symptoms of Hypothyroidism?*

In babies with the inherited form of hypothyroidism, the condition affects growth and cognitive development. It may cause mental retardation, delayed puberty, stunted growth, and ataxia (inability to coordinate muscle movements).

In adults, hypothyroidism slows the body's metabolism, making the patient feel mentally and physically sluggish. Symptoms may include weakness, fatigue, muscle aches, mood swings, hair loss, memory loss, or slow speech. A person's symptoms will depend upon how little thyroid hormone they produce, and for how long they have had the disorder. When the body is deprived of thyroid hormone, the pituitary gland works overtime, producing extra thyroid-stimulating hormone (TSH). This glut of TSH may enlarge the thyroid into a condition called a goiter.

### *How do Doctors Diagnose Hypothyroidism?*

Babies are normally screened for hypothyroidism 24 hours after birth. A tiny sample of blood taken from the baby's heel is tested for low thyroid hormone levels or high thyroid-stimulating hormone (TSH) levels.

### *How is Hypothyroidism Treated?*

Hormone replacement therapy: people with hypothyroidism must take a synthetic form of thyroid hormone every day to reduce their symptoms.

### *Interesting Facts about Hypothyroidism*

One in 5,000 people has hypothyroidism.

## Colon Cancer

### *What is Colon Cancer?*

Cells normally grow and divide only when they are needed to keep our bodies functioning properly. But sometimes, the mechanisms that regulate cell growth stop working and cells divide out of control to form tumors. This is called cancer. When cancer develops in the cells lining the colon (the first part of the large intestine), it is called colon cancer. People who have a history of colon cancer in their family are at greater risk of getting the disease themselves. The risk increases when a relative has had the disease before age 50. These families are considered high-risk, because they may have inherited one of two rare genetic conditions: FAP (familial adenomatous polyposis) or HNPCC (hereditary non-polyposis colon cancer). FAP is caused by mutations of the APC (adenomatous polyposis coli) gene on chromosome 5. APC is a tumour suppressor gene, which means that it prevents uncontrolled cell growth. People who inherit a mutated form of this gene develop growths called polyps in their colon. By age 15, they may have hundreds of these polyps. Polyps are not cancerous at first, but if they aren't treated, they will develop into colon cancer. HNPCC (also called Lynch syndrome) is caused by mutations in one of several genes that fix damaged DNA. People who inherit one of these mutations have a much greater risk of accumulating mutations that will lead to uncontrolled cell growth and cancer.

### How do People get Colon Cancer?

FAP and HNPCC are both inherited in an autosomal dominant pattern. If a parent has FAP or HNPCC, his or her children run a 50 percent risk of inheriting the mutated gene. Usually when a person inherits a defective gene it does not necessarily mean he or she will develop a malignant cancer. However, the APC gene strikingly predisposes one to colon cancer. People who inherit one bad copy of the APC gene are practically guaranteed to develop colon cancer by age 40. Similarly, people who inherit one bad copy of a gene associated with HNPCC have an 80 percent chance of getting colon cancer before. HNPCC also increases a person's risk of developing other cancers, including ovarian, stomach, brain, and liver.

### What are the Symptoms of Colon Cancer?

Colon cancer affects the stomach and bowels. Common symptoms include: diarrhea or constipation, blood in the stool, vomiting, bloating, cramps, and unexplained weight loss.

### How do Doctors Diagnose Colon Cancer?

When a patient shows symptoms of colon cancer, his or her doctor can screen for the disease using one of several tests:

- Fecal Occult Blood Test (FOBT) - Colon cancer can sometimes cause tiny dots of blood, too small for the eye to see, in the feces The FOBT test uses a special chemical to check the patient's stool sample for these traces of blood.

- Flexible-Sigmoidoscopy - Using a thin flexible tube called a simoidoscope, the doctor looks inside the patient's colon for growths called polyps.
- Double Contrast Barium Enema (DCBA) - A silvery-white metallic substance called barium is inserted up the patient's colon through the rectum. The barium outlines the patient's colon on an x-ray screen.
- Colonoscopy - Using a thin instrument called a colonoscope, the doctor looks inside the patient's colon. During the procedure, the doctor removes pieces of tissue (called a biopsy) to test them for cancer. If the doctor finds any polyps, he or she can also remove them. A newer method, called virtual colonoscopy, looks at the colon without going into the body, with an MRI or CT scan.
- DNA-Based stool test - This test examines DNA taken from a patient's stool sample to look for genetic defects associated with colon cancer.

### *How is Colon Cancer Treated?*

Colon cancer is very treatable. In fact, about 90 percent of patients survive the disease after treatment. First, doctors stage the disease to see how far it has progressed. If the cancer has not spread to other tissues of the body, it can be treated with: special chemicals (chemotherapy) or radiation (powerful x-rays) that kill all rapidly dividing cells in the body, including cancer cells surgery to remove the polyps and/or cancerous part of the colon.

## Interesting Facts about Colon Cancer

People who have FAP can develop hundreds and even thousands of polyps in their colon, whereas people with HNPCC develop relatively few. The progression from a benign to a malignant cancer typically requires multiple mutations that allow cells to acquire new and abnormal characteristics, such as an increased growth rate, inability to adhere or stick to neighbouring cells, propensity to migrate to other places in the body, etc. For example, at least seven mutations are required to produce a malignant colon tumour. Inherited cancers often provide clues about the genes mutated in noninherited (sporadic) cancers. For example, mutations in the APC gene are found not only in FAP tumors but in 85% of all sporadic colon tumors as well.

## Breast Cancer and Ovarian Cancer

### *What is Breast Cancer and Ovarian Cancer?*

Cells normally grow and divide only when they are needed to keep our bodies functioning properly. But sometimes, the mechanisms that regulate cell growth stop working and cells divide out of control to form tumors. This is called cancer. When cancer develops in the cells of breast or ovarian tissue it is called breast or ovarian cancer, respectively. Most people who develop breast or ovarian cancer have no history of the disease in their family. In fact, only 5 to 10 percent of all breast and ovarian cancers are caused by inherited genetic factors. These rare cases typically result from inherited mutations in either the BRCA1 or BRCA2 gene.

The BRCA1 and BRCA2 genes are called tumour suppressor genes because they prevent uncontrolled cell growth. BRCA1 is located on chromosome 17, and BRCA2 is located on chromosome 13. Scientists believe BRCA1 and BRCA2 work by fixing damaged or broken DNA. Women who inherit a mutated copy of the BRCA1 or BRCA2 gene accumulate broken and deformed chromosomes, and therefore have a greater chance of accumulating mutations that will lead to uncontrolled cell growth and cancer. Men who inherit the defective genes are also more likely to develop breast and/or prostrate cancer. (Yes, men can get breast cancer.)

### *How do you get Breast or Ovarian Cancers?*

High-risk families include those who have inherited a mutation in either the BRCA1 or BRCA2 gene. The mutated BRCA1 and BRCA2 genes are inherited in an autosomal dominant pattern. A child only needs to inherit one copy of the mutated gene to have an increased cancer risk. Children who have a parent with the BRCA1 or BRCA2 mutation have a 50 percent chance of inheriting the mutation. Just because a person inherits the defective gene does not mean he or she will develop cancer, but inheritance greatly increases the risk. Out of every 100 women who inherit a mutated BRCA1 or BRCA2 gene, as many as 60 will develop breast cancer by age 50; by age 70, approximately 80 will develop breast cancer.

### *How do Doctors Test for BRCA1 and BRCA2 Mutations?*

A person with a strong family history of breast or ovarian cancer is a likely candidate for genetic screening. By analyzing a sample of the patient's blood, doctors can identify whether the person has inherited the BRCA1 or BRCA2 mutation.

The test cannot tell if or when the person will develop breast or ovarian cancer; it can only tell if he or she is at risk because of the faulty gene.

### *What can a Woman do if she Inherits these Mutations?*

Some women who discover that they've inherited the mutated BRCA1 and BRCA2 genes undergo special treatments to protect themselves from breast and ovarian cancer.

They may have one or both breasts removed (called a mastectomy or double mastectomy), and their ovaries removed.

They may take the medicine tamoxifen, which is believed to protect against breast cancer.

## Interesting Facts about Breast and Ovarian Cancer

The progression from a benign to a malignant cancer typically requires multiple mutations that allow cells to acquire new and abnormal characteristics, such as an increased growth rate, inability to adhere or stick to neighbouring cells, propensity to migrate to other places in the body, etc. Genes involved in the repair of DNA damage (such as BRCA1 and BRCA2) are often associated with cancer.

This is because they allow mutations to accumulate at a much faster rate.

Usually, people inherit one of several hundred different mutations of the BRCA1 and BRCA2 genes. But for some reason, Eastern European (Ashkenazi) Jews seem to inherit only three of theses different mutations. Ashkenazi Jews are also 10 times more likely to have mutations in the BRCA1 and BRCA2 genes than any other ethnic group.

## Alzheimer's Disease

### What is Alzheimer's Disease?

Alzheimer's is a disease that causes dementia, or loss of brain function. It affects the parts of the brain that deal with memory, thought, and language. The brain of a person with Alzheimer's contains abnormal clumps of cellular debris and protein(plaques) and collapsed microtubules (support structures of the cell). Microtubule disintegration is caused by a malfunctioning protein called tau, which normally stabalizes the microtubules. In Alzheimer's patients, tau proteins instead cluster together to form disabling tangles. These plaques and tangles damage the healthy cells around them. The brain also produces smaller amounts of neurotransmitters (acetylcholine, serotonin, and norepinephrine), chemicals that allow nerve cells to talk to one another.

The most common form of the disease, which strikes after age 65, is linked to the apolipoprotein E (apoE) gene on chromosome 19. Scientists don't know how apoE4 increases the risk of developing Alzheimer's. They do know that everyone has apoE, which comes in three forms. One of the forms (apoE4) increases a person's risk of developing Alzheimer's. The other two forms seem to protect against the disease.

While people who inherit the apoE4 form of the gene are at increased risk for the disease, they will not necessarily develop it. Mutations in genes found on chromosomes 1, 14, and 21 are linked to rarer forms of the disease, which strike earlier in life.

### How do People get Alzheimer's Disease?

Scientists don't know exactly how people develop Alzheimer's, but they believe it is caused by a combination of genes and environmental factors (multifactorial disorder). The early-onset forms of Alzheimer's are inherited in an autosomal dominant pattern, which means that only one parent has to pass down a defective copy of the gene for their child to develop the disorder.

### What are the Symptoms of Alzheimer's Disease?

Because Alzheimer's destroys brain cells, people who have the disorder slowly lose their ability to think clearly. At first, they may forget words or names, or have trouble finding things. As the disorder worsens, they may forget how to do simple tasks (such as walking to a friend's house or brushing their hair). Some people with Alzheimer's also feel nervous or sad.

### *How do Doctors Diagnose Alzheimer's Disease?*

There is no one test for Alzheimer's. Doctors use several different tests to check a patient's memory, language skills, and problem solving abilities. These tests don't diagnose Alzheimer's, but they can rule out other disorders that have similar symptoms.

### *How is Alzheimer's Disease Treated?*

There is no cure for Alzheimer's, but a few medicines can slow its symptoms. A drug called Aricept increases the amount of the neurotransmitter acetylcholine in the brain. A new medicine, Namenda, protects brain cells from a chemical called glutamate, which can damage nerve cells.

Doctors may also give their Alzheimer's patients antidepressants or anti-anxiety medicines to ease some of their symptoms. People with Alzheimer's often need a caregiver - someone to help them get around and do the things they were once able to do themselves.

### *Interesting Facts about Alzheimer's Disease*

Alzheimer's was named after the German doctor, Alois Alzheimer, who first named the disorder in 1906. The older a person gets, the higher his or her risk of getting Alzheimer's. Only about 1 or 2 people out of 100 have Alzheimer's at age 65; whereas, one out of every five people has the disorder by age 80.

## Down Syndrome

### *What is Down Syndrome?*

Down syndrome is a developmental disorder caused by an extra copy of chromosome 21 (which is why the disorder is also called "trisomy 21"). Having an extra copy of this chromosome means that each gene may be producing more protein product than normal. Cells seem to tolerate this better than having not enough protein, or having altered protein due to a mutation in the DNA sequence. However, producing too much protein can also have serious consequences, as seen in Down syndrome. Genes on chromosome 21 that specifically contribute to the various symptoms of Down syndrome are now being identified.

### How do People get Down Syndrome?

Down syndrome is typically caused by what is called nondisjunction. If a pair of number 21 chromosomes fails to separate during the formation of an egg (or sperm), this is referred to as nondisjunction. When that egg unites with a normal sperm to form an embryo, that embryo ends up with three copies of chromosome 21 instead of the normal two. The extra chromosome is then copied in every cell of the baby's body. Interestingly, nondisjunction events seem to occur more frequently in older women. This may explain why the risk of having a baby with Down syndrome is greater among mothers age 35 and older. In rare cases Down syndrome is caused by a Robertsonian translocation, which occurs when the long arm of chromosome 21 breaks off and attaches to another chromosome at the centromere. The carrier of such a translocation will not have Down syndrome, but can produce children with Down syndrome. Javascript is required to view this content.

### What are the Symptoms of Down Syndrome?

People with Down syndrome have very distinct facial features: a flat face, a small broad nose, abnormally shaped ears, a large tongue, and upward slanting eyes with small folds of skin in the corners.

People with Down syndrome have an increased risk of developing a number of medically significant problems: respiratory infections, gastrointestinal tract obstruction (blocked digestive tract), leukemia, heart defects, hearing loss, hypothyroidism, and various eye abnormalities.

They also exhibit moderate to severe mental retardation; children with Down syndrome usually develop more slowly than their peers, and have trouble learning to walk, talk, and take care of themselves. Because of these medical problems most people with Down syndrome have a decreased life expectancy. About half live to be 50 years of age.

### How do Doctors Diagnose Down Syndrome?

Two types of tests check for Down syndrome during a woman's pregnancy: screening and diagnostic tests. Screening tests identify a mother who is likely carrying a baby with Down syndrome. The most common screening tests are the Triple Screen and the Alpha-Fetoprotein Plus. These tests measure levels of certain substances in the blood. Alternatively, ultrasounds (which use sound waves to look inside the mother's uterus) allow the doctor to examine the fetus in the womb for the physical signs of Down syndrome.

To confirm a positive result identified in a screening test, one of the following diagnostic tests can be performed: chorionic villus sampling (CVS), amniocentesis, and percutaneous umbilical blood sampling (PUBS).

Each takes a sample from the placenta, amniotic fluid, or umbilical cord, respectively, to examine the baby's chromosomes and determine if he or she has an extra chromosome 21. If Down syndrome is not diagnosed in the womb, doctors can usually recognize it after the baby is born by the distinctive facial features. The diagnosis is confirmed with a karyotype - an examination of the baby's chromosomes.

### How is Down Syndrome Treated?

No cure exists for Down syndrome. But physical therapy and/or speech therapy can help people with the disorder develop more normally. Screening for common medical problems associated with the disorder, followed by corrective surgery, can often improve quality of life. Moreover, enriched environments significantly increase their capacity to learn and lead a meaningful life.

### Interesting facts about Down syndrome

Down syndrome is really the only trisomy compatible with life. Only two other trisomies have been observed in babies born alive (trisomies 13 and 18), but babies born with these trisomies have only a 5% chance of surviving longer than one year. In 90% of Trisomy 21 cases, the additional chromosome comes from the mother's egg rather than the father's sperm. Down syndrome is the most common genetic disorder caused by a chromosomal abnormality. It affects 1 out of every 800 to

1,000 babies. Down syndrome was originally described in 1866 by John Langdon Down. It wasn't until 1959 that a French doctor, named Jerome Lejeune, discovered it was caused by the inheritance of an extra chromosome 21.

## Severe Combined Immunodeficiency

### *What is Severe Combined Immunodeficiency (SCID)?*

SCID is a group of very rare-and potentially fatal-inherited disorders related to the immune system. The immune system normally fights off attacks from dangerous bacteria and viruses. People with SCID have a defect in their immune system that leaves them vulnerable to potentially deadly infections. There are several types of SCID. The most common form is caused by a mutation in the SCIDX1 gene located on the X chromosome. This gene encodes a protein that is used to construct a receptor called IL2RG (interleukin-2 receptor).

These receptors reside in the plasma membrane of immune cells. Their job is to allow two types of immune cells - T cells and B cells - to communicate. When the gene is mutated, the receptors cannot form and are absent from immune cells. As a result, the immune cells can't communicate with one another about invaders in the environment. Not enough T and B cells are produced to fight off the infection, and the body is left defenseless. Another form of SCID is caused by a mutation on chromosome 20 and is characterized by a deficiency of the enzyme adenosine deaminase.

### *How do people get SCID?*

The most common form of SCID exhibits an X-linked recessive pattern of inheritance, and is therefore referred to as X-linked SCID. When a gene is located on the X chromosome, males are more often affected than females. Males do not have a second X chromosome to compensate for the defective one. They only need to inherit one bad copy of the gene to have the symptoms of the disorder.

Females, on the other hand, have two X chromosomes. If they inherit one defective X chromosome, they still have its healthy pair. They don't develop symptoms of the disorder, but they still carry the faulty gene and can pass it on to their children.

### *What are the Symptoms of SCID?*

Symptoms usually appear in the first few months of life. Because the immune system cannot protect the baby's body, babies with the disorder tend to get one infection after another. Some of these bacterial infections may be life-threatening, including pneumonia (lung infection), meningitis (brain infection), and sepsis (blood infection).

To make matters worse, SCID patients often don't respond to the antibiotics used to treat bacterial infections. They may suffer more frequently from ear infections, sinus infections, a chronic cough, and rashes on the skin.

Early diagnosis of SCID is very important, because without quick treatment, children with the disease aren't likely to live past age 2.

# Chapter 8

# Organisation of Genome

Genome organization refers to the sequential, not the structural organization of the genome. Besides the coding exons, the non-coding DNA in Eukaryotes may fall in the following classes:

- Introns. They are DNA sequences inserted between the exons and found in the ORF.
- Pseudogenes. 'Dead', non-functional copies of genes present elsewhere in the genome, but no longer of any use.
- Retropseudogenes. Like pseudogenes, but have been processed, i.e. lack introns. Produced by the action of reverse transcriptase (RT) on mRNA, and subsequent incorporation of the cDNA into the genome.
- Transposons. Jumping genes, which splice themselves in and out of the genome (in DNA form) randomly, by the action of transposase.
- Retrotransposons. Transcribed into an mRNA, which encodes an RT enzyme, which then copies the mRNA back to DNA and incorporates it into the genome.

Infact in humans only 1.5% of the entire genome length corresponds to coding DNA. This 1.5% codes for about 27,000 genes which in turn code for proteins that are responsible for all the cellular processes.

## Genome Analysis?

Genome analysis entails the prediction of genes in uncharacterized genomic sequences. The 21st century has seen the announcement of the draft version of the human genome sequence. Model organisms have been sequenced in both the plant and animal kingdoms. However, the pace of genome annotation is not matching the pace of genome sequencing. Experimental genome annotation is slow and time consuming. The demand is to be able to develop computational tools for gene prediction.

Computational Gene prediction is relatively simple for the prokaryotes where all the genes are converted into the corresponding mRNA and then into proteins. The process is more complex for eukaryotic cells where the coding DNA sequence is interrupted by random sequences called introns.

Some of the questions which biologists want to answer today are:

- Given a DNA sequence, what part of it codes for a protein and what part of it is junk DNA.
- Classify the junk DNA as intron, untranslated region, transposons, dead genes, regulatory elements etc.
- Divide a newly sequenced genome into the genes (coding) and the non-coding regions.

The importance of genome analysis can be understood by comparing the human and chimpanzee genomes. The chimp and human genomes vary by an average of just 2% i.e. just about 160 enzymes.

A complete genome analysis of the two genomes would give a strong insight into the various mechanisms responsible for the differences.

Arabidopsis and Humans have the same number of genes, though the Arabidopsis genome is around 250 times smaller than humans. How is that ?

The human genome has a lot of junk DNA, specifically transposons and mobile genetic elements. This increases the size of the human genome, though the number of genes is only 27,000.

However, the number of protein products in humans is significantly higher. Many of the sequenced human genes have alternative splice products. In addition, several other processes (e.g. signal transduction) proceed via further protein modifications, such as Glycosylation. Therefore, the number of human protein products could far exceed the number of genes. Why do plants have such bulky genomes when they are not as complex as some of the higher eukaryotes ? This is mostly due to two factors: the ability of plants to duplicate their genomes in order to reproduce (a process known as polyploidization) and the susceptibility of plants to mobile genetic elements.

- Polyploidization allows plants to more easily form hybrids when pollen and ova from different species come together. The result of such hybridization events are plants with genomes that are the sum of the two parent genome sizes (as opposed to half of one parent's genome and half of the other parent's genome as in normal sexual reproduction.
- Also, in case of plants, it is fairly common to observe insertion of transposable elements in intergenic regions. This also explains the difference in the sizes of plant genomes among themselves as well.

## What is Open Reading Frame (ORF) ?

The region of the nucleotide sequences from the start codon (ATG) to the stop codon is called the Open Reading frame. Gene finding in organism specially prokaryotes starts form searching for an open reading frames (ORF). An ORF is a sequence of DNA that starts with start codon "ATG" (not always) and ends with any of the three termination codons (TAA, TAG, TGA). Depending on the starting point, there are six possible ways (three on forward strand and three on complementary strand) of translating any nucleotide sequence into amino acid sequence according to the genetic code.These are called reading frames. While eukaryotic gene finding is altogether a different task as the eukaryotic genes are not continuous and interrupted by intervening noncoding sequences called 'introns'. Moreover organization of genetic information in eukaryotes and prokaryotes is different.

## What is Coding Sequence(CDS)? How is it Different from the ORF?

The Coding Sequence (CDS) is the actual region of DNA that is translated to form proteins. While the ORF may contain introns as well, the CDS refers to those nucleotides(concatenated exons) that can be divided into codons which are actually translated into amino acids by the ribosomal translation machinery. In Prokaryotes the ORF and the CDS are the same.

## DNA Replication

Before a cell divides, its DNA is replicated (duplicated.) Because the two strands of a DNA molecule have complementary base pairs, the nucleotide sequence of each strand automatically supplies the information needed to produce its partner.

If the two strands of a DNA molecule are separated, each can be used as a pattern or template to produce a complementary strand. Each template and its new complement together then form a new DNA double helix, identical to the original.

Before replication can occur, the length of the DNA double helix about to be copied must be unwound.

In addition, the two strands must be separated, much like the two sides of a zipper, by breaking the weak hydrogen bonds that link the paired bases. Once the DNA strands have been unwound, they must be held apart to expose the bases so that new nucleotide partners can hydrogen-bond to them.

The enzyme DNA polymerase then moves along the exposed DNA strand, joining newly arrived nucleotides into a new DNA strand that is complementary to the template.

Each cell contains a family of more than thirty enzymes to insure the accurate replication of DNA.

### *Primers*

Though DNA polymerase can elongate a polynucleotide strand by adding new nucleotides, it cannot start a strand from scratch because it can only bond new nucleotides to a free sugar (3') end of a nucleotide chain. DNA polymerase requires the assistance of a primer, a previously existing short strand of DNA (or RNA) that is complementary to the first part of the DNA segment being copied.

This small strand of nucleotides anneals (binds) by complementary base pairing to the beginning of the area being copied. With the primer in place, DNA polymerase is then able to continue adding the rest of the pairs of the segment until a new double strand of DNA is completed. Primers are formed from free nucleotides in the cell by enzymes called DNA primases.

## Replication Occurs Differently on Antiparallel Strands of DNA

That nucleotides can be added only to the sugar or 3' end of the growing complementary chain presents no problem for the side of the DNA chain opening at its phosphate or 5' end. The primer that binds to the first few exposed bases will end with a sugar (3') where the phosphate of a new nucleotide can be attached. From there on, DNA polymerase can continuously synthesize the growing complementary strand. This strand of DNA is called the leading strand. A different challenge faces DNA polymerase when the complementary side of the DNA molecule begins unzipping from its sugar (3') toward its phosphate (5') end. A primer of complementary molecules attaching to the opening end of this chain would have a phosphate not a sugar at its exposed end so that new nucleotides could not be joined. To get around this problem, this strand is synthesized in small pieces backward from the overall direction of replication. This strand is called the lagging strand. The short segments of newly assembled DNA from which the lagging strand is built are called Okazaki fragments. As replication proceeds and nucleotides are added to the 3' end of the Okazaki fragments, they come to meet each other. The primer fragments are then booted out by enzymes and replaced by appropriate DNA nucleotides. The whole thing is then stitched together by another enzyme called DNA ligase.

## DNA Damage, Repair, and Recombination

DNA repair refers to a collection of processes by which a cell identifies and corrects damage to the DNA molecules that encode its genome. In human cells, both normal metabolic activities and environmental factors such as UV light and radiation can cause DNA damage, resulting in as many as 1 million individual molecular lesions per cell per day.

Many of these lesions cause structural damage to the DNA molecule and can alter or eliminate the cell's ability to transcribe the gene that the affected DNA encodes. Other lesions induce potentially harmful mutations in the cell's genome, which affect the survival of its daughter cells after it undergoes mitosis. As a consequence, the DNA repair process is constantly active as it responds to damage in the DNA structure. When normal repair processes fail, and when cellular apoptosis does not occur, irreparable DNA damage may occur, including double-strand breaks and DNA crosslinkages.

The rate of DNA repair is dependent on many factors, including the cell type, the age of the cell, and the extracellular environment. A cell that has accumulated a large amount of DNA damage, or one that no longer effectively repairs damage incurred to its DNA, can enter one of three possible states:

1. an irreversible state of dormancy, known as senescence
2. cell suicide, also known as apoptosis or programmed cell death
3. unregulated cell division, which can lead to the formation of a tumor that is cancerous.

The DNA repair ability of a cell is vital to the integrity of its genome and thus to its normal functioning and that of the organism. Many genes that were initially shown to influence life span have turned out to be involved in DNA damage repair and protection. Failure to correct molecular lesions in cells that form gametes can introduce mutations into the genomes of the offspring and thus influence the rate of evolution.

## DNA Damage

DNA damage, due to environmental factors and normal metabolic processes inside the cell, occurs at a rate of 1,000 to 1,000,000 molecular lesions per cell per day. While this constitutes only 0.000165% of the human genome's approximately 6 billion bases (3 billion base pairs), unrepaired lesions in critical genes (such as tumor suppressor genes) can impede a cell's ability to carry out its function and appreciably increase the likelihood of tumour formation.

The vast majority of DNA damage affects the primary structure of the double helix; that is, the bases themselves are chemically modified. These modifications can in turn disrupt the molecules' regular helical structure by introducing non-native chemical bonds or bulky adducts that do not fit in the standard double helix. Unlike proteins and RNA, DNA usually lacks tertiary structure and therefore damage or disturbance does not occur at that level. DNA is, however, supercoiled and wound around "packaging" proteins called histones (in eukaryotes), and both superstructures are vulnerable to the effects of DNA damage.

### *Sources of Damage*

DNA damage can be subdivided into two main types:

1. endogenous damage such as attack by reactive oxygen species produced from normal metabolic byproducts (spontaneous mutation), especially the process of oxidative deamination
    - also includes replication errors.
2. exogenous damage caused by external agents such as
    - ultraviolet [UV 200-300nm] radiation from the sun
    - other radiation frequencies, including x-rays and gamma rays
    - hydrolysis or thermal disruption
    - certain plant toxins
    - human-made mutagenic chemicals, especially aromatic compounds that act as DNA intercalating agents
    - cancer chemotherapy and radiotherapy
    - viruses.

The replication of damaged DNA before cell division can lead to the incorporation of wrong bases opposite damaged ones.

Daughter cells that inherit these wrong bases carry mutations from which the original DNA sequence is unrecoverable (except in the rare case of a back mutation, for example, through gene conversion).

### *Types of Damage*

There are five main types of damage to DNA due to endogenous cellular processes:

1. *oxidation* of bases [e.g. 8-oxo-7,8-dihydroguanine (8-oxoG)] and generation of DNA strand interruptions from reactive oxygen species,

2. *alkylation* of bases (usually methylation), such as formation of 7-methylguanine, 1-methyladenine, 6-O-Methylguanine
3. *hydrolysis* of bases, such as deamination, depurination, and depyrimidination.
4. "bulky adduct formation" (i.e., benzo[a]pyrene diol epoxide-dG adduct)
5. *mismatch* of bases, due to errors in DNA replication, in which the wrong DNA base is stitched into place in a newly forming DNA strand, or a DNA base is skipped over or mistakenly inserted.

Damage caused by exogenous agents comes in many forms. Some examples are:

1. *UV-B light* causes crosslinking between adjacent cytosine and thymine bases creating *pyrimidine dimers*. This is called direct DNA damage.
2. *UV-A light* creates mostly free radicals. The damage caused by free radicals is called indirect DNA damage.
3. *Ionizing radiation* such as that created by radioactive decay or in *cosmic rays* causes breaks in DNA strands. Low-level ionizing radiation may induce irreparable DNA damage (leading to replicational and transcriptional errors needed for neoplasia or may trigger viral interactions) leading to pre-mature aging and cancer.
4. *Thermal disruption* at elevated temperature increases the rate of depurination (loss of purine bases from the DNA backbone) and single-strand breaks. For example, hydrolytic depurination is seen in the thermophilic bacteria, which grow in hot springs at 40-80 °C. The rate of depurination (300 purine residues per genome per generation) is too high in these species to be repaired by normal repair machinery, hence a possibility of an adaptive response cannot be ruled out.
5. *Industrial chemicals* such as vinyl chloride and hydrogen peroxide, and environmental chemicals such as polycyclic hydrocarbons found in smoke, soot and tar create a huge diversity of DNA adducts- ethenobases,

oxidized bases, alkylated phosphotriesters and Crosslinking of DNA just to name a few.

UV damage, alkylation/methylation, X-ray damage and oxidative damage are examples of induced damage. Spontaneous damage can include the loss of a base, deamination, sugar ring puckering and tautomeric shift.

### *Nuclear versus Mitochondrial DNA Damage*

In human cells, and eukaryotic cells in general, DNA is found in two cellular locations - inside the nucleus and inside the mitochondria. Nuclear DNA (nDNA) exists as chromatin during non-replicative stages of the cell cycle and is condensed into aggregate structures known as chromosomes during cell division. In either state the DNA is highly compacted and wound up around bead-like proteins called histones.

Whenever a cell needs to express the genetic information encoded in its nDNA the required chromosomal region is unravelled, genes located therein are expressed, and then the region is condensed back to its resting conformation. Mitochondrial DNA (mtDNA) is located inside mitochondria organelles, exists in multiple copies, and is also tightly associated with a number of proteins to form a complex known as the nucleoid.

Inside mitochondria, reactive oxygen species (ROS), or free radicals, byproducts of the constant production of adenosine triphosphate (ATP) via oxidative phosphorylation, create a highly oxidative environment that is known to damage mtDNA. A critical enzyme in counteracting the toxicity of these species is superoxide dismutase, which is present in both the mitochondria and cytoplasm of eukaryotic cells.

### *Senescence and Apoptosis*

Senescence, an irreversible state in which the cell no longer divides, is a protective response to the shortening of the chromosome ends. The telomeres are long regions of repetitive noncoding DNA that cap chromosomes and undergo partial degradation each time a cell undergoes division. In contrast, quiescence is a reversible state of cellular dormancy that is unrelated to genome damage. Senescence in cells may serve as a functional alternative to apoptosis in cases where the physical presence of a cell for spatial reasons is required by the organism, which serves as a "last resort" mechanism to prevent a cell with damaged DNA from replicating inappropriately in the absence of pro-growth cellular signaling. Unregulated cell division can lead to the formation of a tumour, which is potentially lethal to an organism. Therefore, the induction of senescence and apoptosis is considered to be part of a strategy of protection against cancer.

### DNA Damage and Mutation

It is important to distinguish between DNA damage and mutation, the two major types of error in DNA. DNA damages and mutation are fundamentally different. Damages are physical abnormalities in the DNA, such as single- and double-strand breaks, 8-hydroxydeoxyguanosine residues, and polycyclic aromatic hydrocarbon adducts.

DNA damages can be recognized by enzymes, and, thus, they can be correctly repaired if redundant information, such as the undamaged sequence in the complementary DNA strand or in a homologous chromosome, is available for copying. If a cell retains DNA damage, transcription of a gene can be prevented, and, thus, translation into a protein will also be blocked. Replication may also be blocked and/or the cell may die.

In contrast to DNA damage, a mutation is a change in the base sequence of the DNA. A mutation cannot be recognized by enzymes once the base change is present in both DNA strands, and, thus, a mutation cannot be repaired. At the cellular level, mutations can cause alterations in protein function and regulation. Mutations are replicated when the cell replicates. In a population of cells, mutant cells will increase or decrease in frequency according to the effects of the mutation on the ability of the cell to survive and reproduce. Although distinctly different from each other, DNA damages and mutations are related because DNA damages often cause errors of DNA synthesis during replication or repair; these errors are a major source of mutation.

Given these properties of DNA damage and mutation, it can be seen that DNA damages are a special problem in non-dividing or slowly dividing cells, where unrepaired damages will tend to accumulate over time. On the other hand, in rapidly dividing cells, unrepaired DNA damages that do not kill the cell by blocking replication will tend to cause replication errors and thus mutation. The great majority of mutations that are not neutral in their effect are deleterious to a cell's survival. Thus, in a population of cells comprising a tissue with replicating cells, mutant cells will tend to be lost. However, infrequent mutations that provide a survival advantage will tend to clonally expand at the expense of neighboring cells in the tissue. This advantage to the cell is disadvantageous to the whole organism, because such mutant cells can give rise to cancer. Thus, DNA damages in frequently dividing cells, because they give rise to mutations, are a prominent cause of cancer. In contrast, DNA damages in infrequently dividing cells are likely a prominent cause of aging.

### DNA Repair Mechanisms

Cells cannot function if DNA damage corrupts the integrity and accessibility of essential information in the genome (but cells remain superficially functional when so-called "non-essential" genes are missing or damaged). Depending on the type of damage inflicted on the DNA's double helical structure, a variety of repair strategies have evolved to restore lost information. If possible, cells use the unmodified complementary strand of the DNA or the sister chromatid as a template to recover the original information. Without access to a template, cells use an error-prone recovery mechanism known as translesion synthesis as a last resort. Damage to DNA alters the spatial configuration of the helix, and such alterations can be detected by the cell. Once damage is localized, specific DNA repair molecules bind at or near the site of damage, inducing other molecules to bind and form a complex that enables the actual repair to take place.

### Direct Reversal

Cells are known to eliminate three types of damage to their DNA by chemically reversing it. These mechanisms do not require a template, since the types of damage they counteract can occur in only one of the four bases. Such direct reversal mechanisms are specific to the type of damage incurred and do not involve breakage of the phosphodiester backbone.

The formation of pyrimidine dimers upon irradiation with UV light results in an abnormal covalent bond between adjacent pyrimidine bases. The photoreactivation process directly reverses this damage by the action of the enzyme photolyase, whose activation is obligately dependent on energy absorbed from blue/UV light (300–500 nm wavelength) to promote catalysis. Another type of damage, methylation of guanine bases, is directly reversed by the protein methyl guanine methyl transferase (MGMT), the bacterial equivalent of which is called ogt. This is an expensive process because each MGMT molecule can be used only once; that is, the reaction is stoichiometric rather than catalytic.

A generalized response to methylating agents in bacteria is known as the adaptive response and confers a level of resistance to alkylating agents upon sustained exposure by upregulation of alkylation repair enzymes. The third type of DNA damage reversed by cells is certain methylation of the bases cytosine and adenine.

### *Single-strand Damage*

When only one of the two strands of a double helix has a defect, the other strand can be used as a template to guide the correction of the damaged strand. In order to repair damage to one of the two paired molecules of DNA, there exist a number of excision repair mechanisms that remove the damaged nucleotide and replace it with an undamaged nucleotide complementary to that found in the undamaged DNA strand.

1. Base excision repair (BER), which repairs damage to a single base caused by oxidation, alkylation, hydrolysis, or deamination. The damaged base is removed by a DNA glycosylase. The "missing tooth" is then recognised by an enzyme called AP endonuclease, which cuts the Phosphodiester bond. The missing part is then resynthesized by a DNA polymerase, and a DNA ligase performs the final nick-sealing step.
2. Nucleotide excision repair (NER), which recognizes bulky, helix-distorting lesions such as pyrimidine dimers and 6,4 photoproducts. A specialized form of NER known as transcription-coupled repair deploys NER enzymes to genes that are being actively transcribed.

3. Mismatch repair (MMR), which corrects errors of DNA replication and recombination that result in mispaired (but undamaged) nucleotides.

### *Double-strand Breaks*

Double-strand breaks, in which both strands in the double helix are severed, are particularly hazardous to the cell because they can lead to genome rearrangements. Three mechanisms exist to repair double-strand breaks (DSBs): non-homologous end joining (NHEJ), microhomology-mediated end joining (MMEJ), and homologous recombination. PVN Acharya noted that double-strand breaks and a "cross-linkage joining both strands at the same point is irreparable because neither strand can then serve as a template for repair. The cell will die in the next mitosis or in some rare instances, mutate."

In NHEJ, DNA Ligase IV, a specialized DNA ligase that forms a complex with the cofactor XRCC4, directly joins the two ends. To guide accurate repair, NHEJ relies on short homologous sequences called microhomologies present on the single-stranded tails of the DNA ends to be joined. If these overhangs are compatible, repair is usually accurate. NHEJ can also introduce mutations during repair. Loss of damaged nucleotides at the break site can lead to deletions, and joining of nonmatching termini forms translocations. NHEJ is especially important before the cell has replicated its DNA, since there is no template available for repair by homologous recombination. There are "backup" NHEJ pathways in higher eukaryotes. Besides its role as a genome caretaker, NHEJ is required for joining hairpin-capped double-strand breaks induced during V(D)J recombination, the process that generates diversity in B-cell and T-cell receptors in the vertebrate immune system.

Homologous recombination requires the presence of an identical or nearly identical sequence to be used as a template for repair of the break. The enzymatic machinery responsible for this repair process is nearly identical to the machinery responsible for chromosomal crossover during meiosis. This pathway allows a damaged chromosome to be repaired using a sister chromatid (available in G2 after DNA replication) or a homologous chromosome as a template. DSBs caused by the replication machinery attempting to synthesize across a single-strand break or unrepaired lesion cause collapse of the replication fork and are typically repaired by recombination.

Topoisomerases introduce both single- and double-strand breaks in the course of changing the DNA's state of supercoiling, which is especially common in regions near an open replication fork. Such breaks are not considered DNA damage because they are a natural intermediate in the topoisomerase biochemical mechanism and are immediately repaired by the enzymes that created them.

A team of French researchers bombarded *Deinococcus radiodurans* to study the mechanism of double-strand break DNA repair in that organism. At least two copies of the genome, with random DNA breaks, can form DNA fragments through annealing. Partially overlapping fragments are then used for synthesis of homologous regions through a moving D-loop that can continue extension until they find complementary partner strands. In the final step there is crossover by means of RecA-dependent homologous recombination.

### *Translesion Synthesis*

Translesion synthesis is a DNA damage tolerance process that allows the DNA replication machinery to replicate past DNA lesions such as thymine dimers or AP sites. It involves switching out regular DNA polymerases for specialized translesion polymerases (i.e. DNA polymerase IV or V, from the Y Polymerase family), often with larger active sites that can facilitate the insertion of bases opposite damaged nucleotides.

The polymerase switching is thought to be mediated by, among other factors, the post-translational modification of the replication processivity factor PCNA. Translesion synthesis polymerases often have low fidelity (high propensity to insert wrong bases) on undamaged templates relative to regular polymerases. However, many are extremely efficient at inserting correct bases opposite specific types of damage. For example, Pol ç mediates error-free bypass of lesions induced by UV irradiation, whereas Pol æ introduces mutations at these sites. From a cellular perspective, risking the introduction of point mutations during translesion synthesis may be preferable to resorting to more drastic mechanisms of DNA repair, which may cause gross chromosomal aberrations or cell death. In short, the process involves specialized polymerases either bypassing or repairing lesions at locations of stalled DNA replication. A bypass platform is provided to these polymerases by Proliferating cell nuclear antigen (PCNA). Under normal circumstances, PCNA bound to polymerases replicates the DNA. At a site of lesion, PCNA is ubiquitinated, or modified, by the RAD6/RAD18 proteins to provide a platform for the specialized polymerases to bypass the lesion and resume DNA replication.

## Global Response to DNA Damage

Cells exposed to ionizing radiation, ultraviolet light or chemicals are prone to acquire multiple sites of bulky DNA lesions and double-strand breaks. Moreover, DNA damaging agents can damage other biomolecules such as proteins, carbohydrates, lipids, and RNA. The accumulation of damage, to be specific, double-strand breaks or adducts stalling the replication forks, are among known stimulation signals for a global response to DNA damage. The global response to damage is an act directed toward the cells' own preservation and triggers multiple pathways of macromolecular repair, lesion bypass, tolerance, or apoptosis. The common features of global response are induction of multiple genes, cell cycle arrest, and inhibition of cell division.

### *DNA Damage Checkpoints*

After DNA damage, cell cycle checkpoints are activated. Checkpoint activation pauses the cell cycle and gives the cell time to repair the damage before continuing to divide. DNA damage checkpoints occur at the G1/S and G2/M boundaries. An intra-S checkpoint also exists. Checkpoint activation is controlled by two master kinases, ATM and ATR. ATM responds to DNA double-strand breaks and disruptions in chromatin structure, whereas ATR primarily responds to stalled replication forks. These kinases phosphorylate downstream targets in a signal transduction cascade, eventually leading to cell cycle arrest. A class of checkpoint mediator proteins including BRCA1, MDC1, and 53BP1 has also been identified. These proteins seem to be required for transmitting the checkpoint activation signal to downstream proteins.

P53 is an important downstream target of ATM and ATR, as it is required for inducing apoptosis following DNA damage. At the G1/S checkpoint, p53 functions by deactivating the CDK2/cyclin E complex. Similarly, p21 mediates the G2/M checkpoint by deactivating the CDK1/cyclin B complex.

### The Prokaryotic SOS Response

The SOS response is the term used to describe changes in gene expression in *Escherichia coli* and other bacteria in response to extensive DNA damage. The prokaryotic SOS system is regulated by two key proteins: LexA and RecA. The LexA homodimer is a transcriptional repressor that binds to operator sequences commonly referred to as SOS boxes.

In *Escherichia coli* it is known that LexA regulates transcription of approximately 48 genes including the lexA and recA genes. The SOS response is known to be widespread in the Bacteria domain, but it is mostly absent in some bacterial phyla, like the Spirochetes.

The most common cellular signals activating the SOS response are regions of single-stranded DNA (ssDNA), arising from stalled replication forks or double-strand breaks, which are processed by DNA helicase to separate the two DNA strands. In the initiation step, RecA protein binds to ssDNA in an ATP hydrolysis driven reaction creating RecA–ssDNA filaments. RecA–ssDNA filaments activate LexA autoprotease activity, which ultimately leads to cleavage of LexA dimer and subsequent LexA degradation. The loss of LexA repressor induces transcription of the SOS genes and allows for further signal induction, inhibition of cell division and an increase in levels of proteins responsible for damage processing.

In *Escherichia coli*, SOS boxes are 20-nucleotide long sequences near promoters with palindromic structure and a high degree of sequence conservation. In other classes and phyla, the sequence of SOS boxes varies considerably, with different length and composition, but it is always highly conserved and one of the strongest short signals in the genome. The high information content of SOS boxes permits differential binding of LexA to different promoters and allows for timing of the SOS response. The lesion repair genes are induced at the beginning of SOS response. The error-prone translesion polymerases, for example, UmuCD'2 (also called DNA polymerase V), are induced later on as a last resort. Once the DNA damage is repaired or bypassed using polymerases or through recombination, the amount of single-stranded DNA in cells is decreased, lowering the amounts of RecA filaments decreases cleavage activity of LexA homodimer, which then binds to the SOS boxes near promoters and restores normal gene expression.

### *Eukaryotic Transcriptional Responses to DNA Damage*

Eukaryotic cells exposed to DNA damaging agents also activate important defensive pathways by inducing multiple proteins involved in DNA repair, cell cycle checkpoint control, protein trafficking and degradation. Such genome wide transcriptional response is very complex and tightly regulated, thus allowing coordinated global response to damage. Exposure of yeast *Saccharomyces cerevisiae* to DNA damaging agents results in overlapping but distinct transcriptional profiles. Similarities to environmental shock response indicates that a general global stress response pathway exist at the level of transcriptional activation. In contrast, different human cell types respond to damage differently indicating an absence of a common global response. The probable explanation for this difference between yeast and human cells may be in the heterogeneity of mammalian cells. In an animal different types of cells are distributed among different organs that have evolved different sensitivities to DNA damage. In general global response to DNA damage involves expression of multiple genes responsible for postreplication repair, homologous recombination, nucleotide excision repair, DNA damage checkpoint, global transcriptional activation, genes controlling mRNA decay, and many others. A large amount of damage to a cell leaves it with an important decision: undergo apoptosis and die, or survive at the cost of living with a modified genome. An increase in tolerance to damage can lead to an increased rate of survival that will allow a greater accumulation of mutations. Yeast Rev1 and human polymerase ç are members of [Y family translesion DNA polymerases present during global response to DNA damage and are responsible for enhanced mutagenesis during a global response to DNA damage in eukaryotes.

## DNA Repair and Aging

### *Pathological Effects of Poor DNA Repair*

DNA repair rate is an important determinant of cell pathology.

Experimental animals with genetic deficiencies in DNA repair often show decreased life span and increased cancer incidence. For example, mice deficient in the dominant NHEJ pathway and in telomere maintenance mechanisms get lymphoma and infections more often, and, as a consequence, have shorter lifespans than wild-type mice.

In similar manner, mice deficient in a key repair and transcription protein that unwinds DNA helices have premature onset of aging-related diseases and consequent shortening of lifespan. However, not every DNA repair deficiency creates exactly the predicted effects; mice deficient in the NER pathway exhibited shortened life span without correspondingly higher rates of mutation.

If the rate of DNA damage exceeds the capacity of the cell to repair it, the accumulation of errors can overwhelm the cell and result in early senescence, apoptosis, or cancer. Inherited diseases associated with faulty DNA repair functioning result in premature aging, increased sensitivity to carcinogens, and correspondingly increased cancer risk.

On the other hand, organisms with enhanced DNA repair systems, such as *Deinococcus radiodurans*, the most radiation-resistant known organism, exhibit remarkable resistance to the double-strand break-inducing effects of radioactivity, likely due to enhanced efficiency of DNA repair and especially NHEJ.

### *Longevity and Caloric Restriction*

A number of individual genes have been identified as influencing variations in life span within a population of organisms. The effects of these genes is strongly dependent on the environment, in particular, on the organism's diet. Caloric restriction reproducibly results in extended lifespan in a variety of organisms, likely via nutrient sensing pathways and decreased metabolic rate. The molecular mechanisms by which such restriction results in lengthened lifespan are as yet unclear; however, the behavior of many genes known to be involved in

DNA repair is altered under conditions of caloric restriction. For example, increasing the gene dosage of the gene SIR-2, which regulates DNA packaging in the nematode worm *Caenorhabditis elegans*, can significantly extend lifespan. The mammalian homolog of SIR-2 is known to induce downstream DNA repair factors involved in NHEJ, an activity that is especially promoted under conditions of caloric restriction. Caloric restriction has been closely linked to the rate of base excision repair in the nuclear DNA of rodents, although similar effects have not been observed in mitochondrial DNA.

It is interesting to note that the *C. elegans* gene AGE-1, an upstream effector of DNA repair pathways, confers dramatically extended life span under free-feeding conditions but leads to a decrease in reproductive fitness under conditions of caloric restriction. This observation supports the pleiotropy theory of the biological origins of aging, which suggests that genes conferring a large survival advantage early in life will be selected for even if they carry a corresponding disadvantage late in life.

### Medicine and DNA Repair Modulation

***Hereditary DNA Repair Disorders:*** Defects in the NER mechanism are responsible for several genetic disorders, including:

- Xeroderma pigmentosum: hypersensitivity to sunlight/UV, resulting in increased skin cancer incidence and premature aging
- Cockayne syndrome: hypersensitivity to UV and chemical agents
- Trichothiodystrophy: sensitive skin, brittle hair and nails.

Mental retardation often accompanies the latter two disorders, suggesting increased vulnerability of developmental neurons.

Other DNA repair disorders include:

- Werner's syndrome: premature aging and retarded growth
- Bloom's syndrome: sunlight hypersensitivity, high incidence of malignancies (especially leukemias).

- Ataxia telangiectasia: sensitivity to ionizing radiation and some chemical agents.

All of the above diseases are often called "segmental progerias" ("accelerated aging diseases") because their victims appear elderly and suffer from aging-related diseases at an abnormally young age, while not manifesting all the symptoms of old age.

Other diseases associated with reduced DNA repair function include Fanconi's anemia, hereditary breast cancer and hereditary colon cancer.

### *DNA Repair and Cancer*

Inherited mutations that affect DNA repair genes are strongly associated with high cancer risks in humans. Hereditary nonpolyposis colorectal cancer (HNPCC) is strongly associated with specific mutations in the DNA mismatch repair pathway. BRCA1 and BRCA2, two famous mutations conferring a hugely increased risk of breast cancer on carriers, are both associated with a large number of DNA repair pathways, especially NHEJ and homologous recombination.

Cancer therapy procedures such as chemotherapy and radiotherapy work by overwhelming the capacity of the cell to repair DNA damage, resulting in cell death. Cells that are most rapidly dividing - most typically cancer cells - are preferentially affected. The side-effect is that other non-cancerous but rapidly dividing cells such as stem cells in the bone marrow are also affected. Modern cancer treatments attempt to localize the DNA damage to cells and tissues only associated with cancer, either by physical means (concentrating the therapeutic agent in the region of the tumour) or by biochemical means (exploiting a feature unique to cancer cells in the body).

## DNA Repair and Evolution

The basic processes of DNA repair are highly conserved among both prokaryotes and eukaryotes and even among bacteriophage (viruses that infect bacteria); however, more complex organisms with more complex genomes have correspondingly more complex repair mechanisms. The ability of a large number of protein structural motifs to catalyze relevant chemical reactions has played a significant role in the elaboration of repair mechanisms during evolution. For an extremely detailed review of hypotheses relating to the evolution of DNA repair.

The fossil record indicates that single-cell life began to proliferate on the planet at some point during the Precambrian period, although exactly when recognizably modern life first emerged is unclear. Nucleic acids became the sole and universal means of encoding genetic information, requiring DNA repair mechanisms that in their basic form have been inherited by all extant life forms from their common ancestor. The emergence of Earth's oxygen-rich atmosphere (known as the "oxygen catastrophe") due to photosynthetic organisms, as well as the presence of potentially damaging free radicals in the cell due to oxidative phosphorylation, necessitated the evolution of DNA repair mechanisms that act specifically to counter the types of damage induced by oxidative stress.

### *Rate of Evolutionary Change*

On some occasions, DNA damage is not repaired, or is repaired by an error-prone mechanism that results in a change from the original sequence. When this occurs, mutations may propagate into the genomes of the cell's progeny. Should such an event occur in a germ line cell that will eventually produce a gamete, the mutation has the potential to be passed on to the organism's offspring. The rate of evolution in a particular species (or, in a particular gene) is a function of the rate of mutation. As a consequence, the rate and accuracy of DNA repair mechanisms have an influence over the process of evolutionary change.

## Senescence

Senescence or biological aging is the change in the biology of an organism as it ages after its maturity. Such changes range from those affecting its cells and their function to that of the whole organism. There are a number of theories as to why senescence occurs, including ones that claim it is programmed by gene expression changes and that it is the accumulative damage of biological processes.

The word *senescence* is derived from the Latin word *senex*, meaning *old man*, *old age*, or *advanced in age*.

### Cellular Senescence

*Cellular senescence* is the phenomenon by which normal diploid cells lose the ability to divide, normally after about 50 cell divisions in vitro. Some cells become senescent after fewer replications cycles as a result of DNA double strand breaks, toxins, etc. This phenomenon is also known as "replicative senescence", the "Hayflick phenomenon", or the Hayflick limit in honour of Dr. Leonard Hayflick who was the first to publish this information in 1965. In response to DNA damage (including shortened telomeres), cells either age or self-destruct (apoptosis, programmed cell death) if the damage cannot be repaired. In this 'cellular suicide', the death of one cell, or more, may benefit the organism as a whole. For example, in plants the death of the water-conducting xylem cells (tracheids and vessel elements) allows the cells to function more efficiently and so deliver water to the upper parts of a plant.

### Aging of the Whole Organism

*Organismal senescence* is the aging of whole organisms. In general, aging is characterized by the declining ability to respond to stress, increased homeostatic imbalance, and increased risk of aging-associated diseases. Death is the ultimate consequence of aging, though "old age" is not a scientifically recognized cause of death because there is always a specific proximal cause, such as cancer, heart disease, or liver failure. Aging of whole organisms is therefore a complex process that can be defined as "a progressive deterioration of physiological function, an intrinsic age-related process of loss of viability and increase in vulnerability". Differences in maximum life span among species correspond to different "rates of aging". For example, inherited differences in the rate of aging make a mouse elderly at 3 years and a human elderly at 80 years. These genetic differences affect a variety of physiological processes, including the efficiency of DNA repair, antioxidant enzymes, and rates of free radical production.

Senescence of the organism gives rise to the Gompertz–Makeham law of mortality, which says that mortality rate rises rapidly with age.

Some animals, such as some reptiles and fish, age slowly (negligible senescence) and exhibit very long lifespans. Some even exhibit "negative senescence", in which mortality falls with age, in disagreement with the Gompertz–Makeham "law".

Whether replicative senescence (Hayflick limit) plays a causative role in organismal aging is at present an active area of investigation.

### *Theories of Aging*

The process of senescence is complex, and may derive from a variety of different mechanisms and exist for a variety of different reasons. However, senescence is not universal, and scientific evidence suggests that cellular senescence evolved in certain species because it prevents the onset of cancer. In a few simple species, such as Hydra, senescence is negligible and cannot be detected. All such species have no "post-mitotic" cells; they reduce the effect of damaging free radicals by cell division and dilution. Such species are not immortal, however, as they will eventually fall prey to trauma or disease. Moreover, average lifespans can vary greatly within and between species. This suggests that both genetic and environmental factors contribute to aging.

In general, theories that explain senescence have been divided between the programmed and stochastic theories of aging. Programmed theories imply that aging is regulated by biological clocks operating throughout the lifespan. This regulation would depend on changes in gene expression that affect the systems responsible for maintenance, repair, and defense responses. Stochastic theories blame environmental impacts on living organisms that induce cumulative damage at various levels as the cause of aging, examples of which ranging from damage to DNA, damage to tissues and cells by oxygen radicals (widely known as free radicals countered by the even more well-known antioxidants), and cross-linking.

However, aging is seen as a progressive failure of homeodynamics (homeostasis) involving genes for the maintenance and repair, stochastic events leading to molecular damage and molecular heterogeneity, and chance events determining the probability of death. Since complex and interacting systems of maintenance and repair comprise the homeodynamic (old term: homeostasis) space of a biological system, aging is considered to be a progressive shrinkage of homeodynamic space mainly due to increased molecular heterogeneity.

### *Evolutionary Theories*

A gene can be expressed at various life-stages. Therefore, natural selection can support lethal and harmful alleles, if their expression occurs after reproduction. Senescence may be the product of such selection. In addition, aging is believed to have evolved because of the increasingly smaller probability of an organism still being alive at older age, due to predation and accidents, both of which may be random and age-invariant. It is thought that strategies that result in a higher reproductive rate at a young age, but shorter overall lifespan, result in a higher lifetime reproductive success and are therefore favoured by natural selection.

In essence, aging is, therefore, the result of investing resources in reproduction, rather than maintenance of the body (the "Disposable Soma" theory), in light of the fact that accidents, predation, and disease will eventually kill the organism no matter how much energy is devoted to repair of the body. Various other theories of aging exist, and are not necessarily mutually exclusive.

The geneticist J. B. S. Haldane wondered why the dominant mutation that causes Huntington's disease remained in the population, and why natural selection had not eliminated it. The onset of this neurological disease is (on average) at age 45 and is invariably fatal within 10–20 years.

Haldane assumed that, in human prehistory, few survived until age 45. Since few were alive at older ages and their contribution to the next generation was therefore small relative to the large cohorts of younger age groups, the force of selection against such late-acting deleterious mutations was correspondingly small.

However, if a mutation affected younger individuals, selection against it would be strong. Therefore, late-acting deleterious mutations could accumulate in populations over evolutionary time through genetic drift. This principle has been demonstrated experimentally. And it is these later-acting deleterious mutations that are believed to cause—even allow—age-related mortality.

Peter Medawar formalised this observation in his mutation accumulation theory of aging. "The force of natural selection weakens with increasing age—even in a theoretically immortal population, provided only that it is exposed to real hazards of mortality. If a genetic disaster... happens late enough in individual life, its consequences may be completely unimportant". The 'real hazards of mortality' are, in typical circumstances, predation, disease, and accidents. So, even an immortal population, whose fertility does not decline with time, will have fewer individuals alive in older age groups. This is called 'extrinsic mortality'. Young cohorts, not depleted in numbers yet by extrinsic mortality, contribute far more to the next generation than the few remaining older cohorts, so the force of selection against late-acting deleterious mutations, which affect only these few older individuals, is very weak. The mutations may not be selected against, therefore, and may spread over evolutionary time into the population.

The major testable prediction made by this model is that species that have high extrinsic mortality in nature will age more quickly and have shorter intrinsic lifespans. This is borne out among mammals, the best-studied in terms of life history. There is a correlation among mammals between body size and lifespan, such that larger species live longer than smaller species in controlled/optimum conditions, but there are notable exceptions. For instance, many bats and rodents are of similar size, yet bats live much longer. For instance, the little brown bat, half the size of a mouse, can live 30 years in the wild. A mouse will live 2–3 years even with optimum conditions. The explanation is that bats have fewer predators, so therefore low extrinsic mortality. Thus, more individuals survive to later ages, so the force of selection against late-acting deleterious mutations is stronger. Fewer late-acting deleterious mutations = slower aging = longer lifespan. Birds are also warm-blooded and are of similar size to many small mammals, yet live often 5–10 times as long. They have fewer predation pressures compared with ground-dwelling mammals. And seabirds, which, in general, have the fewest predators of all birds, live longest.

Also, when examining the body-size vs. lifespan relationship, one will observe that predator mammals tend to have longer lifespans than prey animals in a controlled environment such as a zoo or nature reserve. The explanation for the long lifespans of primates (such as humans, monkeys, and apes) relative to body size is that their intelligence and often sociality helps them avoid becoming prey. Being a predator, being smart, and working together all reduce extrinsic mortality.

Another evolutionary theory of aging was proposed by George C. Williams and involves antagonistic pleiotropy. A single gene may affect multiple traits. Some traits that increase fitness early in life may also have negative effects later in life. But, because many more individuals are alive at young ages than at old ages, even small positive effects early can be strongly selected for, and large negative effects later may be very weakly selected against.

Williams suggested the following example: Perhaps a gene codes for calcium deposition in bones, which promotes juvenile survival and will therefore be favored by natural selection; however, this same gene promotes calcium deposition in the arteries, causing negative effects in old age. Therefore, negative effects in old age may reflect the result of natural selection for pleiotropic genes that are beneficial early in life. In this case, fitness is relatively high when Fisher's reproductive value is high and relatively low when Fisher's reproductive value is low.

### *Gene Regulation*

A number of genetic components of aging have been identified using model organisms, ranging from the simple budding yeast *Saccharomyces cerevisiae* to worms such as *Caenorhabditis elegans* and fruit flies (*Drosophila melanogaster*). Study of these organisms has revealed the presence of at least two conserved aging pathways.

One of these pathways involves the gene *Sir2*, a NAD+-dependent histone deacetylase. In yeast, Sir2 is required for genomic silencing at three loci: The yeast mating loci, the telomeres and the ribosomal DNA (rDNA). In some species of yeast, replicative aging may be partially caused by homologous recombination between rDNA repeats; excision of rDNA repeats results in the formation of extrachromosomal rDNA circles (ERCs). These ERCs replicate and preferentially segregate to the mother cell during cell division, and are believed to result in cellular senescence by titrating away (competing for) essential nuclear factors. ERCs have not been observed in other species (nor even all strains of the same yeast species) of yeast (which also display replicative senescence), and ERCs are not believed to contribute to aging in higher organisms such as humans (they have not been shown to accumulate in mammals in a similar manner to yeast).

Extrachromosomal circular DNA (eccDNA) has been found in worms, flies, and humans. The origin and role of eccDNA in aging, if any, is unknown.

Despite the lack of a connection between circular DNA and aging in higher organisms, extra copies of Sir2 are capable of extending the lifespan of both worms and flies (though, in flies, this finding has not been replicated by other investigators, and the activator of Sir2 resveratrol does not reproducibly increase lifespan in either species).

Whether the Sir2 homologues in higher organisms have any role in lifespan is unclear, but the human SIRT1 protein has been demonstrated to deacetylate p53, Ku70, and the forkhead family of transcription factors. SIRT1 can also regulate acetylates such as CBP/p300, and has been shown to deacetylate specific histone residues.

RAS1 and RAS2 also affect aging in yeast and have a human homologue. RAS2 overexpression has been shown to extend lifespan in yeast. Other genes regulate aging in yeast by increasing the resistance to oxidative stress. Superoxide dismutase, a protein that protects against the effects of mitochondrial free radicals, can extend yeast lifespan in stationary phase when overexpressed.

In higher organisms, aging is likely to be regulated in part through the insulin/IGF-1 pathway. Mutations that affect insulin-like signaling in worms, flies, and the growth hormone/IGF1 axis in mice are associated with extended lifespan.

In yeast, Sir2 activity is regulated by the nicotinamidase PNC1. PNC1 is transcriptionally upregulated under stressful conditions such as caloric restriction, heat shock, and osmotic shock. By converting nicotinamide to niacin, nicotinamide is removed, inhibiting the activity of Sir2. A nicotinamidase found in humans, known as PBEF, may serve a similar function, and a secreted form of PBEF known as visfatin may help to regulate serum insulin levels. It is not known, however, whether these mechanisms also exist in humans, since there are obvious differences in biology between humans and model organisms.

Sir2 activity has been shown to increase under calorie restriction. Due to the lack of available glucose in the cells, more NAD+ is available and can activate Sir2. Resveratrol, a stilbenoid found in the skin of red grapes, was reported to extend the lifespan of yeast, worms, and flies (the lifespan extension in flies and worms have proved irreproducible by independent investigators). It has been shown to activate Sir2 and therefore mimics the effects of calorie restriction, if one accepts that caloric restriction is indeed dependent on Sir2.

Gene expression is imperfectly controlled, and it is possible that random fluctuations in the expression levels of many genes contribute to the aging process as suggested by a study of such genes in yeast. Individual cells, which are genetically identical, none-the-less can have substantially different responses to outside stimuli, and markedly different lifespans, indicating the epigenetic factors play an important role in gene expression and aging as well as genetic factors.

According to the GenAge database of aging-related genes there are over 700 genes associated with aging in model organisms: 555 in the soil roundworm (*Caenorhabditis elegans*), 87 in the bakers' yeast (*Saccharomyces cerevisiae*), 75 in the fruit fly (*Drosophila melanogaster*) and 68 in the mouse (*Mus musculus*).

### Cellular Senescence

As noted above, senescence is not universal, and senescence is not observed in single-celled organisms that reproduce through the process of cellular mitosis. Moreover, cellular senescence is not observed in several organisms, including perennial plants, sponges, corals, and lobsters. In those species where cellular senescence is observed, cells eventually become post-mitotic when they can no longer replicate themselves through the process of cellular mitosis; i.e., cells experience *replicative senescence*. How and why some cells become post-mitotic in some species has been the subject of much research and speculation, but (as noted above) it is widely believed that cellular senescence evolved as a way to prevent the onset and spread of cancer. Somatic cells that have divided many times will have accumulated DNA mutations and would therefore be in danger of becoming cancerous if cell division continued.

Lately, the role of telomeres in cellular senescence has aroused general interest, especially with a view to the possible genetically adverse effects of cloning. The successive shortening of the chromosomal telomeres with each cell cycle is also believed to limit the number of divisions of the cell, thus contributing to aging. There have, on the other hand, also been reports that cloning could alter the shortening of telomeres. Some cells do not age and are, therefore, described as being "biologically immortal". It is theorized by some that when it is discovered exactly what allows these cells, whether it be the result of telomere lengthening or not, to divide without limit that it will be possible to genetically alter other cells to have the same capability. It is further theorized that it will eventually be possible to genetically engineer all cells in the human body to have this capability by employing gene therapy and, therefore, stop or reverse aging, effectively making the entire organism potentially immortal.

Cancer cells are usually immortal. In about 85% of tumors, this evasion of cellular senescence is the result of up-activation of their telomerase genes. This simple observation suggests that reactivation of telomerase in healthy individuals could greatly increase their cancer risk. Whether cell senescence plays any role in organismal aging is at present unknown, and is an active area of investigation. Mouse mutants lacking telomerase do not immediately show accelerated aging.

### Chemical Damage

One of the earliest aging theories was the *Rate of Living Hypothesis* described by Raymond Pearl in 1928(based on earlier work by Max Rubner), which states that fast basal metabolic rate corresponds to short maximum life span.

While there may be some validity to the idea that for various types of specific damage detailed below that are by-products of metabolism, all other things being equal, a fast metabolism may reduce lifespan, in general this theory does not adequately explain the differences in lifespan either within, or between, species. Calorically-restricted animals process as much, or more, calories per gram of body mass, as their *ad libitum* fed counterparts, yet exhibit substantially longer lifespans.. Similarly, metabolic rate is a poor predictor of lifespan for birds, bats and other species that, it is presumed, have reduced mortality from predation, and therefore have evolved long lifespans even in the presence of very high metabolic rates. More recently, it was shown that, when modern statistical methods for correcting for the effects of body size and phylogeny are employed, metabolic rate does not correlate with longevity in mammals or birds.

With respect to specific types of chemical damage caused by metabolism, it is suggested that damage to long-lived biopolymers, such as structural proteins or DNA, caused by ubiquitous chemical agents in the body such as oxygen and sugars, are in part responsible for aging. The damage can include breakage of biopolymer chains, cross-linking of biopolymers, or chemical attachment of unnatural substituents (haptens) to biopolymers.

Under normal aerobic conditions, approximately 4% of the oxygen metabolized by mitochondria is converted to superoxide ion, which can subsequently be converted to hydrogen peroxide, hydroxyl radical and eventually other reactive species including other peroxides and singlet oxygen, which can, in turn, generate free radicals capable of damaging structural proteins and DNA. Certain metal ions found in the body, such as copper and iron, may participate in the process. (In Wilson's disease, a hereditary defect that causes the body to retain copper, some of the symptoms resemble accelerated senescence.) These processes are termed *oxidative damage* and are linked to the benefits of nutritionally derived polyphenol antioxidants. Sugars such as glucose and fructose can react with certain amino acids such as lysine and arginine and certain DNA bases such as guanine to produce sugar adducts, in a process called *glycation*. These adducts can further rearrange to form reactive species, which can then cross-link the structural proteins or DNA to similar biopolymers or other biomolecules such as non-structural proteins. People with diabetes, who have elevated blood sugar, develop senescence-associated disorders much earlier than the general population, but can delay such disorders by rigorous control of their blood sugar levels. There is evidence that sugar damage is linked to oxidant damage in a process termed *glycoxidation*.

Free radicals can damage proteins, lipids or DNA. Glycation mainly damages proteins. Damaged proteins and lipids accumulate in lysosomes as lipofuscin. Chemical damage to structural proteins can lead to loss of function; for example, damage to collagen of blood vessel walls can lead to vessel-wall stiffness and, thus, hypertension, and vessel wall thickening and reactive tissue formation (atherosclerosis); similar processes in the kidney can lead to renal failure. Damage to enzymes reduces cellular functionality. Lipid peroxidation of the inner mitochondrial membrane reduces the electric potential and the ability to generate energy. It is probably no accident that nearly all of the so-called "accelerated aging diseases" are due to defective DNA repair enzymes.

It is believed that the impact of alcohol on aging can be partly explained by alcohol's activation of the HPA axis, which stimulates glucocorticoid secretion, long-term exposure to which produces symptoms of aging.

### Reliability Theory

Reliability theory suggests that biological systems start their adult life with a high load of initial damage. Reliability theory is a general theory about systems failure. It allows researchers to predict the age-related failure kinetics for a system of given architecture (reliability structure) and given reliability of its components. Reliability theory predicts that even those systems that composed entirely of non-aging elements (with a constant failure rate) will nevertheless deteriorate (fail more often) with age, if these systems are redundant in irreplaceable elements. Aging, therefore, is a direct consequence of systems.

Reliability theory also predicts the late-life mortality deceleration with subsequent leveling-off, as well as the late-life mortality plateaus, as an inevitable consequence of redundancy exhaustion at extreme old ages. The theory explains why mortality rates increase exponentially with age (the Gompertz law) in many species, by taking into account the initial flaws (defects) in newly formed systems. It also explains why organisms "prefer" to die according to the Gompertz law, while technical devices usually fail according to the Weibull (power) law. Reliability theory allows to specify conditions when organisms die according to the Weibull distribution: Organisms should be relatively free of initial flaws and defects.

The theory makes it possible to find a general failure law applicable to all adult and extreme old ages, where the Gompertz and the Weibull laws are just special cases of this more general failure law. The theory explains why relative differences in mortality rates of compared populations (within a given species) vanish with age (compensation law of mortality), and mortality convergence is observed due to the exhaustion of initial differences in redundancy levels.

### *Miscellaneous*

Recently, a kind of early senescence has been alleged to be a possible unintended outcome of early cloning experiments. The issue was raised in the case of Dolly the sheep, following her death from a contagious lung disease.

The claim that Dolly's early death involved premature senescence has been vigorously contested, and Dolly's creator, Dr. Ian Wilmut has expressed the view that her illness and death were probably unrelated to the fact that she was a clone.

A set of rare hereditary (genetic) disorders, each called progeria, has been known for some time. Sufferers exhibit symptoms resembling accelerated aging, including wrinkled skin. The cause of Hutchinson–Gilford progeria syndrome was reported in the journal *Nature* in May 2003. This report suggests that DNA damage, not oxidative stress, is the cause of this form of accelerated aging.

## Transcription

Transcription is the process of creating a complementary RNA copy of a sequence of DNA. Both RNA and DNA are nucleic acids, which use base pairs of nucleotides as a complementary language that can be converted back and forth from DNA to RNA by the action of the correct enzymes. During transcription, a DNA sequence is read by RNA polymerase, which produces a complementary, antiparallel RNA strand. As opposed to DNA replication, transcription results in an RNA complement that includes uracil (U) in all instances where thymine (T) would have occurred in a DNA complement.

Transcription can be explained easily in 4 or 5 simple steps, each moving like a wave along the DNA.

1. RNA polymerase unwinds/"unzips" the DNA by breaking the hydrogen bonds between complimentary nucleotides.
2. RNA nucleotides are paired with complementary DNA bases.
3. RNA sugar-phosphate backbone forms with assistance from RNA polymerase.
4. Hydrogen bonds of the untwisted RNA+DNA helix break, freeing the newly synthesized RNA strand.
5. If the cell has a nucleus, the RNA is further processed and then moves through the small nuclear pores to the cytoplasm.

Transcription is the first step leading to gene expression. The stretch of DNA transcribed into an RNA molecule is called a *transcription unit* and encodes at least one gene. If the gene transcribed encodes a protein, the result of transcription is messenger RNA (mRNA), which will then be used to create that protein via the process of translation. Alternatively, the transcribed gene may encode for either ribosomal RNA (rRNA) or transfer RNA (tRNA), other components of the protein-assembly process, or other ribozymes.

A DNA transcription unit encoding for a protein contains not only the sequence that will eventually be directly translated into the protein (the *coding sequence*) but also *regulatory sequences* that direct and regulate the synthesis of that protein. The regulatory sequence before (upstream from) the coding sequence is called the five prime untranslated region (5'UTR), and the sequence following (downstream from) the coding sequence is called the three prime untranslated region (3'UTR).

Transcription has some proofreading mechanisms, but they are fewer and less effective than the controls for copying DNA; therefore, transcription has a lower copying fidelity than DNA replication.

As in DNA replication, DNA is read from 3' → 5' during transcription. Meanwhile, the complementary RNA is created from the 5' '! 3' direction. This means its 5' end is created first in base pairing. Although DNA is arranged as two antiparallel strands in a double helix, only one of the two DNA strands, called the template strand, is used for transcription. This is because RNA is only single-stranded, as opposed to double-stranded DNA. The other DNA strand is called the coding strand, because its sequence is the same as the newly created RNA transcript (except for the substitution of uracil for thymine). The use of only the 3' '! 5' strand eliminates the need for the Okazaki fragments seen in DNA replication.

Transcription is divided into 5 stages: *pre-initiation, initiation, promoter clearance, elongation* and *termination.*

## Major Steps

### *Pre-initiation*

In eukaryotes, RNA polymerase, and therefore the initiation of transcription, requires the presence of a core promoter sequence in the DNA. Promoters are regions of DNA that promote transcription and, in eukaryotes, are found at -30, -75, and -90 base pairs upstream from the start site of transcription. Core promoters are sequences within the promoter that are essential for transcription initiation. RNA polymerase is able to bind to core promoters in the presence of various specific transcription factors.

The most common type of core promoter in eukaryotes is a short DNA sequence known as a TATA box, found 25-30 base pairs upstream from the start site of transcription. The TATA box, as a core promoter, is the binding site for a transcription factor known as TATA-binding protein (TBP), which is itself a subunit of another transcription factor, called Transcription Factor II D (TFIID).

After TFIID binds to the TATA box via the TBP, five more transcription factors and RNA polymerase combine around the TATA box in a series of stages to form a preinitiation complex. One transcription factor, DNA helicase, has helicase activity and so is involved in the separating of opposing strands of double-stranded DNA to provide access to a single-stranded DNA template. However, only a low, or basal, rate of transcription is driven by the preinitiation complex alone. Other proteins known as activators and repressors, along with any associated coactivators or corepressors, are responsible for modulating transcription rate. Thus, preinitiation complex contains: 1. Core Promoter Sequence 2. Transcription Factors 3. DNA Helicase 4. RNA Polymerase 5. Activators and Repressors The transcription preinitiation in archaea is, in essence, homologous to that of eukaryotes, but is much less complex. The archaeal preinitiation complex assembles at a TATA-box binding site; however, in archaea, this complex is composed of only RNA polymerase II, TBP, and TFB (the archaeal homologue of eukaryotic transcription factor II B (TFIIB)).

### *Initiation*

In bacteria, transcription begins with the binding of RNA polymerase to the promoter in DNA. RNA polymerase is a core enzyme consisting of five subunits: 2 $\alpha$ subunits, 1 $\beta$ subunit, 1 $\beta$' subunit, and 1 $\omega$ subunit. At the start of initiation, the core enzyme is associated with a sigma factor that aids in finding the appropriate -35 and -10 base pairs downstream of promoter sequences. Transcription initiation is more complex in eukaryotes. Eukaryotic RNA polymerase does not directly recognize the core promoter sequences. Instead, a collection of proteins called transcription factors mediate the binding of RNA polymerase and the initiation of transcription. Only after certain transcription factors are attached to the promoter does the RNA polymerase bind to it. The completed assembly of transcription factors and RNA polymerase bind to the promoter, forming a transcription initiation complex. Transcription in the archaea domain is similar to transcription in eukaryotes.

### *Promoter Clearance*

After the first bond is synthesized, the RNA polymerase must clear the promoter. During this time there is a tendency to release the RNA transcript and produce truncated transcripts. This is called *abortive initiation* and is common for both eukaryotes and prokaryotes. Abortive initiation continues to occur until the $\sigma$ factor rearranges, resulting in the transcription elongation complex (which gives a 35 bp moving footprint). The $\sigma$ factor is released before 80 nucleotides of mRNA are synthesized. Once the transcript reaches approximately 23 nucleotides, it no longer slips and elongation can occur. This, like most of the remainder of transcription, is an energy-dependent process, consuming adenosine triphosphate (ATP).

Promoter clearance coincides with phosphorylation of serine 5 on the carboxy terminal domain of RNA Pol in eukaryotes, which is phosphorylated by TFIIH.

### *Elongation*

One strand of the DNA, the *template strand* (or noncoding strand), is used as a template for RNA synthesis. As transcription proceeds, RNA polymerase traverses the template strand and uses base pairing complementarity with the DNA template to create an RNA copy.

Although RNA polymerase traverses the template strand from 3' '! 5', the coding (non-template) strand and newly-formed RNA can also be used as reference points, so transcription can be described as occurring 5' '! 3'. This produces an RNA molecule from 5' '! 3', an exact copy of the coding strand (except that thymines are replaced with uracils, and the nucleotides are composed of a ribose (5-carbon) sugar where DNA has deoxyribose (one less oxygen atom) in its sugar-phosphate backbone).

Unlike DNA replication, mRNA transcription can involve multiple RNA polymerases on a single DNA template and multiple rounds of transcription (amplification of particular mRNA), so many mRNA molecules can be rapidly produced from a single copy of a gene.

Elongation also involves a proofreading mechanism that can replace incorrectly incorporated bases. In eukaryotes, this may correspond with short pauses during transcription that allow appropriate RNA editing factors to bind. These pauses may be intrinsic to the RNA polymerase or due to chromatin structure.

### *Termination*

Bacteria use two different strategies for transcription termination. In Rho-independent transcription termination, RNA transcription stops when the newly synthesized RNA molecule forms a G-C-rich hairpin loop followed by a run of Us. When the hairpin forms, the mechanical stress breaks the weak rU-dA bonds, now filling the DNA-RNA hybrid. This pulls the poly-U transcript out of the active site of the RNA polymerase, in effect, terminating transcription. In the "Rho-dependent" type of termination, a protein factor called "Rho" destabilizes the interaction between the template and the mRNA, thus releasing the newly synthesized mRNA from the elongation complex.

Transcription termination in eukaryotes is less understood but involves cleavage of the new transcript followed by template-independent addition of *A*s at its new 3' end, in a process called polyadenylation.

### *Measuring and Detecting Transcription*

Transcription can be measured and detected in a variety of ways:

- Nuclear Run-on assay: measures the relative abundance of newly formed transcripts
- RNase protection assay and ChIP-Chip of RNAP: detect active transcription sites
- RT-PCR: measures the absolute abundance of total or nuclear RNA levels, which may however differ from transcription rates
- DNA microarrays: measures the relative abundance of the global total or nuclear RNA levels; however, these may differ from transcription rates
- In situ hybridization: detects the presence of a transcript
- MS2 tagging: by incorporating RNA stem loops, such as MS2, into a gene, these become incorporated into newly synthesized RNA. The stem loops can then be detected using a fusion of GFP and the MS2 coat protein, which has a high affinity, sequence-specific interaction with the MS2 stem loops. The recruitment of GFP to the site of transcription is visualised as a single fluorescent spot. This remarkable new approach has revealed that transcription occurs in discontinuous bursts, or pulses. With the notable exception of in situ techniques, most other methods provide cell population averages, and are not capable of detecting this fundamental property of genes.

- Northern blot: the traditional method, and until the advent of RNA-Seq, the most quantitative
- RNA-Seq: applies next-generation sequencing techniques to sequence whole transcriptomes, which allows the measurement of relative abundance of RNA, as well as the detection of additional variations such as fusion genes, post-translational edits and novel splice sites.

## Transcription Factories

Active transcription units are clustered in the nucleus, in discrete sites called transcription factories or euchromatin. Such sites can be visualized by allowing engaged polymerases to extend their transcripts in tagged precursors (Br-UTP or Br-U) and immuno-labelling the tagged nascent RNA. Transcription factories can also be localized using fluorescence in situ hybridization or marked by antibodies directed against polymerases.

There are ~10,000 factories in the nucleoplasm of a HeLa cell, among which are ~8,000 polymerase II factories and ~2,000 polymerase III factories. Each polymerase II factory contains ~8 polymerases. As most active transcription units are associated with only one polymerase, each factory usually contains ~8 different transcription units. These units might be associated through promoters and/or enhancers, with loops forming a 'cloud' around the factor.

### *History*

A molecule that allows the genetic material to be realized as a protein was first hypothesized by François Jacob and Jacques Monod. RNA synthesis by RNA polymerase was established *in vitro* by several laboratories by 1965; however, the RNA synthesized by these enzymes had properties that suggested the existence of an additional factor needed to terminate transcription correctly.

In 1972, Walter Fiers became the first person to actually prove the existence of the terminating enzyme.

Roger D. Kornberg won the 2006 Nobel Prize in Chemistry "for his studies of the molecular basis of eukaryotic transcription".

### *Reverse Transcription*

Some viruses (such as HIV, the cause of AIDS), have the ability to transcribe RNA into DNA. HIV has an RNA genome that is duplicated into DNA. The resulting DNA can be merged with the DNA genome of the host cell. The main enzyme responsible for synthesis of DNA from an RNA template is called reverse transcriptase. In the case of HIV, reverse transcriptase is responsible for synthesizing a complementary DNA strand (cDNA) to the viral RNA genome. An associated enzyme, ribonuclease H, digests the RNA strand, and reverse transcriptase synthesises a complementary strand of DNA to form a double helix DNA structure. This cDNA is integrated into the host cell's genome via another enzyme (integrase) causing the host cell to generate viral proteins that reassemble into new viral particles. In HIV, subsequent to this, the host cell undergoes programmed cell death, apoptosis of T cells. However, in other retroviruses, the host cell remains intact as the virus buds out of the cell.

Some eukaryotic cells contain an enzyme with reverse transcription activity called telomerase. Telomerase is a reverse transcriptase that lengthens the ends of linear chromosomes. Telomerase carries an RNA template from which it synthesizes DNA repeating sequence, or "junk" DNA. This repeated sequence of DNA is important because, every time a linear chromosome is duplicated, it is shortened in length. With "junk" DNA at the ends of chromosomes, the shortening eliminates some of the non-essential, repeated sequence rather than the protein-encoding DNA sequence farther away from the chromosome end. Telomerase is often activated in cancer cells to enable cancer cells to duplicate their genomes indefinitely without losing important protein-coding DNA sequence. Activation of telomerase could be part of the process that allows cancer cells to become *immortal*. However, the true *in vivo* significance of telomerase has still not been empirically proven.

## Ribonucleic Acid

Ribonucleic acid (RNA) is one of the three major macromolecules (along with DNA and proteins) that are essential for all known forms of life. Like DNA, RNA is made up of a long chain of components called nucleotides. Each nucleotide consists of a nucleobase (sometimes called a nitrogenous base), a ribose sugar, and a phosphate group. The sequence of nucleotides allows RNA to encode genetic information. For example, some viruses use RNA instead of DNA as their genetic material, and all organisms use messenger RNA (mRNA) to carry the genetic information that directs the synthesis of proteins.

Like proteins, some RNA molecules play an active role in cells by catalyzing biological reactions, controlling gene expression, or sensing and communicating responses to cellular signals. One of these active processes is protein synthesis, a universal function whereby mRNA molecules direct the assembly of proteins on ribosomes. This process uses transfer RNA (tRNA) molecules to deliver amino acids to the ribosome, where ribosomal RNA (rRNA) links amino acids together to form proteins.

The chemical structure of RNA is very similar to that of DNA, with two differences—(a) RNA contains the sugar ribose while DNA contains the slightly different sugar deoxyribose (a type of ribose that lacks one oxygen atom), and (b) RNA has the nucleobase uracil while DNA contains thymine (uracil and thymine have similar base-pairing properties). Unlike DNA, most RNA molecules are single-stranded. Single-stranded RNA molecules adopt very complex three-dimensional structures, since they are not restricted to the repetitive double-helical form of double-stranded DNA. RNA is made within living cells by RNA polymerases, enzymes that act to copy a DNA or RNA template into a new RNA strand through processes known as transcription or RNA replication, respectively.

## Comparison with DNA

RNA and DNA are both nucleic acids, but differ in three main ways. First, unlike DNA, which is, in general, double-stranded, RNA is a single-stranded molecule in many of its biological roles and has a much shorter chain of nucleotides. Second, while DNA contains *deoxyribose*, RNA contains *ribose* (in deoxyribose there is no hydroxyl group attached to the pentose ring in the 2' position). These hydroxyl groups make RNA less stable than DNA because it is more prone to hydrolysis. Third, the complementary base to adenine is not thymine, as it is in DNA, but rather uracil, which is an unmethylated form of thymine.

Like DNA, most biologically active RNAs, including mRNA, tRNA, rRNA, snRNAs, and other non-coding RNAs, contain self-complementary sequences that allow parts of the RNA to fold and pair with itself to form double helices. Structural analysis of these RNAs has revealed that they are highly structured. Unlike DNA, their structures do not consist of long double helices but rather collections of short helices packed together into structures akin to proteins. In this fashion, RNAs can achieve chemical catalysis, like enzymes. For instance, determination of the structure of the ribosome—an enzyme that catalyzes peptide bond formation—revealed that its active site is composed entirely of RNA.

### *Structure*

Each nucleotide in RNA contains a ribose sugar, with carbons numbered 1' through 5'. A base is attached to the 1' position, in general, adenine (A), cytosine (C), guanine (G), or uracil (U). Adenine and guanine are purines, cytosine, and uracil are pyrimidines. A phosphate group is attached to the 3' position of one ribose and the 5' position of the next. The phosphate groups have a negative charge each at physiological pH, making RNA a charged molecule (polyanion).

The bases may form hydrogen bonds between cytosine and guanine, between adenine and uracil and between guanine and uracil. However, other interactions are possible, such as a group of adenine bases binding to each other in a bulge, or the GNRA tetraloop that has a guanine–adenine base-pair.

An important structural feature of RNA that distinguishes it from DNA is the presence of a hydroxyl group at the 2' position of the ribose sugar. The presence of this functional group causes the helix to adopt the A-form geometry rather than the B-form most commonly observed in DNA. This results in a very deep and narrow major groove and a shallow and wide minor groove. A second consequence of the presence of the 2'-hydroxyl group is that in conformationally flexible regions of an RNA molecule (that is, not involved in formation of a double helix), it can chemically attack the adjacent phosphodiester bond to cleave the backbone.

RNA is transcribed with only four bases (adenine, cytosine, guanine and uracil), but these bases and attached sugars can be modified in numerous ways as the RNAs mature. Pseudouridine (Ø), in which the linkage between uracil and ribose is changed from a C–N bond to a C–C bond, and ribothymidine (T) are found in various places (the most notable ones being in the TØC loop of tRNA). Another notable modified base is hypoxanthine, a deaminated adenine base whose nucleoside is called inosine (I). Inosine plays a key role in the wobble hypothesis of the genetic code.

There are nearly 100 other naturally occurring modified nucleosides, of which pseudouridine and nucleosides with 2'-O-methylribose are the most common. The specific roles of many of these modifications in RNA are not fully understood. However, it is notable that, in ribosomal RNA, many of the post-transcriptional modifications occur in highly functional regions, such as the peptidyl transferase center and the subunit interface, implying that they are important for normal function.

The functional form of single stranded RNA molecules, just like proteins, frequently requires a specific tertiary structure.

The scaffold for this structure is provided by secondary structural elements that are hydrogen bonds within the molecule. This leads to several recognizable “domains” of secondary structure like hairpin loops, bulges, and internal loops. Since RNA is charged, metal ions such as $Mg^{2+}$ are needed to stabilise many secondary and tertiary structures.

### *Synthesis*

Synthesis of RNA is usually catalyzed by an enzyme—RNA polymerase—using DNA as a template, a process known as transcription. Initiation of transcription begins with the binding of the enzyme to a promoter sequence in the DNA (usually found “upstream” of a gene).

The DNA double helix is unwound by the helicase activity of the enzyme. The enzyme then progresses along the template strand in the 3’ to 5’ direction, synthesizing a complementary RNA molecule with elongation occurring in the 5’ to 3’ direction. The DNA sequence also dictates where termination of RNA synthesis will occur.

RNAs are often modified by enzymes after transcription. For example, a poly(A) tail and a 5' cap are added to eukaryotic pre-mRNA and introns are removed by the spliceosome.

There are also a number of RNA-dependent RNA polymerases that use RNA as their template for synthesis of a new strand of RNA. For instance, a number of RNA viruses (such as poliovirus) use this type of enzyme to replicate their genetic material. Also, RNA-dependent RNA polymerase is part of the RNA interference pathway in many organisms.

### Types of RNA

Messenger RNA (mRNA) is the RNA that carries information from DNA to the ribosome, the sites of protein synthesis (translation) in the cell. The coding sequence of the mRNA determines the amino acid sequence in the protein that is produced. Many RNAs do not code for protein however (about 97% of the transcriptional output is non-protein-coding in eukaryotes).

These so-called non-coding RNAs ("ncRNA") can be encoded by their own genes (RNA genes), but can also derive from mRNA introns. The most prominent examples of non-coding RNAs are transfer RNA (tRNA) and ribosomal RNA (rRNA), both of which are involved in the process of translation.

There are also non-coding RNAs involved in gene regulation, RNA processing and other roles. Certain RNAs are able to catalyse chemical reactions such as cutting and ligating other RNA molecules, and the catalysis of peptide bond formation in the ribosome; these are known as ribozymes.

### In Translation

Messenger RNA (mRNA) carries information about a protein sequence to the ribosomes, the protein synthesis factories in the cell. It is coded so that every three nucleotides (a codon) correspond to one amino acid. In eukaryotic cells, once precursor mRNA (pre-mRNA) has been transcribed from DNA, it is processed to mature mRNA.

This removes its introns—non-coding sections of the pre-mRNA. The mRNA is then exported from the nucleus to the cytoplasm, where it is bound to ribosomes and translated into its corresponding protein form with the help of tRNA. In prokaryotic cells, which do not have nucleus and cytoplasm compartments, mRNA can bind to ribosomes while it is being transcribed from DNA. After a certain amount of time the message degrades into its component nucleotides with the assistance of ribonucleases.

Transfer RNA (tRNA) is a small RNA chain of about 80 nucleotides that transfers a specific amino acid to a growing polypeptide chain at the ribosomal site of protein synthesis during translation. It has sites for amino acid attachment and an anticodon region for codon recognition that binds to a specific sequence on the messenger RNA chain through hydrogen bonding. Ribosomal RNA (rRNA) is the catalytic component of the ribosomes. Eukaryotic ribosomes contain four different rRNA molecules: 18S, 5.8S, 28S and 5S rRNA. Three of the rRNA molecules are synthesized in the nucleolus, and one is synthesized elsewhere. In the cytoplasm, ribosomal RNA and protein combine to form a nucleoprotein called a ribosome. The ribosome binds mRNA and carries out protein synthesis. Several ribosomes may be attached to a single mRNA at any time. rRNA is extremely abundant and makes up 80% of the 10 mg/ml RNA found in a typical eukaryotic cytoplasm.

Transfer-messenger RNA (tmRNA) is found in many bacteria and plastids. It tags proteins encoded by mRNAs that lack stop codons for degradation and prevents the ribosome from stalling.

### *Regulatory RNAs*

Several types of RNA can downregulate gene expression by being complementary to a part of an mRNA or a gene's DNA. MicroRNAs (miRNA; 21-22 nt) are found in eukaryotes and act through RNA interference (RNAi), where an effector complex of miRNA and enzymes can break down mRNA to which the miRNA is complementary, block the mRNA from being translated, or accelerate its degradation. While small interfering RNAs (siRNA; 20-25 nt) are often produced by breakdown of viral RNA, there are also endogenous sources of siRNAs.

siRNAs act through RNA interference in a fashion similar to miRNAs. Some miRNAs and siRNAs can cause genes they target to be methylated, thereby decreasing or increasing transcription of those genes. Animals have Piwi-interacting

RNAs (piRNA; 29-30 nt) which are active in germline cells and are thought to be a defense against transposons and play a role in gametogenesis. Many prokaryotes have CRISPR RNAs, a regulatory system similar to RNA interference. Antisense RNAs are widespread; most downregulate a gene, but a few are activators of transcription. One way antisense RNA can act is by binding to an mRNA, forming double-stranded RNA that is enzymatically degraded.

There are many long noncoding RNAs that regulate genes in eukaryotes, one such RNA is Xist, which coats one X chromosome in female mammals and inactivates it. An mRNA may contain regulatory elements itself, such as riboswitches, in the 5' untranslated region or 3' untranslated region; these cis-regulatory elements regulate the activity of that mRNA. The untranslated regions can also contain elements that regulate other genes.

### *In RNA Processing*

Many RNAs are involved in modifying other RNAs. Introns are spliced out of pre-mRNA by spliceosomes, which contain several small nuclear RNAs (snRNA), or the introns can be ribozymes that are spliced by themselves. RNA can also be altered by having its nucleotides modified to other nucleotides than A, C, G and U.

In eukaryotes, modifications of RNA nucleotides are generally directed by small nucleolar RNAs (snoRNA; 60-300 nt), found in the nucleolus and cajal bodies. snoRNAs associate with enzymes and guide them to a spot on an RNA by basepairing to that RNA. These enzymes then perform the nucleotide modification. rRNAs and tRNAs are extensively modified, but snRNAs and mRNAs can also be the target of base modification.

### *RNA Genomes*

Like DNA, RNA can carry genetic information. RNA viruses have genomes composed of RNA, and a variety of proteins encoded by that genome. The viral genome is replicated by some of those proteins, while other proteins protect the genome as the virus particle moves to a new host cell. Viroids are another group of pathogens, but they consist only of RNA, do not encode any protein and are replicated by a host plant cell's polymerase.

### In Reverse Transcription

Reverse transcribing viruses replicate their genomes by reverse transcribing DNA copies from their RNA; these DNA copies are then transcribed to new RNA. Retrotransposons also spread by copying DNA and RNA from one another, and telomerase contains an RNA that is used as template for building the ends of eukaryotic chromosomes.

### Double-stranded RNA

Double-stranded RNA (dsRNA) is RNA with two complementary strands, similar to the DNA found in all cells. dsRNA forms the genetic material of some viruses (double-stranded RNA viruses). Double-stranded RNA such as viral RNA or siRNA can trigger RNA interference in eukaryotes, as well as interferon response in vertebrates.

## Key Discoveries in RNA Biology

Research on RNA has led to many important biological discoveries and numerous Nobel Prizes. Nucleic acids were discovered in 1868 by Friedrich Miescher, who called the material 'nuclein' since it was found in the nucleus. It was later discovered that prokaryotic cells, which do not have a nucleus, also contain nucleic acids. The role of RNA in protein synthesis was suspected already in 1939. Severo Ochoa won the 1959 Nobel Prize in Medicine (shared with Arthur Kornberg) after he discovered an enzyme that can synthesize RNA in the laboratory. Ironically, the enzyme discovered by Ochoa (polynucleotide phosphorylase) was later shown to be responsible for RNA degradation, not RNA synthesis.

The sequence of the 77 nucleotides of a yeast tRNA was found by Robert W. Holley in 1965, winning Holley the 1968 Nobel Prize in Medicine (shared with Har Gobind Khorana and Marshall Nirenberg). In 1967, Carl Woese hypothesized that RNA might be catalytic and suggested that the earliest forms of life (self-replicating molecules) could have relied on RNA both to carry genetic information and to catalyze biochemical reactions—an RNA world.

During the early 1970s, retroviruses and reverse transcriptase were discovered, showing for the first time that enzymes could copy RNA into DNA (the opposite of the usual route for transmission of genetic information). For this work, David Baltimore, Renato Dulbecco and Howard Temin were awarded a Nobel Prize in 1975. In 1976, Walter Fiers and his team determined the first complete nucleotide sequence of an RNA virus genome, that of bacteriophage MS2.

In 1977, introns and RNA splicing were discovered in both mammalian viruses and in cellular genes, resulting in a 1993 Nobel to Philip Sharp and Richard Roberts. Catalytic RNA molecules (ribozymes) were discovered in the early 1980s, leading to a 1989 Nobel award to Thomas Cech and Sidney Altman. In 1990 it was found in petunia that introduced genes can silence similar genes of the plant's own, now known to be a result of RNA interference.

At about the same time, 22 nt long RNAs, now called microRNAs, were found to have a role in the development of *C. elegans*. Studies on RNA interference gleaned a Nobel Prize for Andrew Fire and Craig Mello in 2006, and another Nobel was awarded for studies on transcription of RNA to Roger Kornberg in the same year. The discovery of gene regulatory RNAs has led to attempts to develop drugs made of RNA, such as siRNA, to silence genes.

## Transcription and Processing of RNA

### Gene Expression: Transcription

The majority of genes are expressed as the proteins they encode. The process occurs in two steps:

- Transcription = DNA '! RNA
- Translation = RNA '! protein.

Taken together, they make up the "central dogma" of biology: DNA '! RNA '! protein.

## Gene Transcription: DNA '! RNA

DNA serves as the template for the synthesis of RNA much as it does for its own replication.

### The Steps

- Some 50 different protein transcription factors bind to promoter sites, usually on the 52 side of the gene to be transcribed.

- An enzyme, an RNA polymerase, binds to the complex of transcription factors.
- Working together, they open the DNA double helix.
- The RNA polymerase proceeds to "read" one strand moving in its 32 '! 52 direction.
- In eukaryotes, this requires — at least for protein-encoding genes — that the nucleosomes in front of the advancing RNA polymerase (RNAP II) be removed. A complex of proteins is responsible for this. The same complex replaces the nucleosomes after the DNA has been transcribed and RNAP II has moved on.
- As the RNA polymerase travels along the DNA strand, it assembles ribonucleotides (supplied as triphosphates, e.g., ATP) into a strand of RNA.
- Each ribonucleotide is inserted into the growing RNA strand following the rules of base pairing. Thus for each C encountered on the DNA strand, a G is inserted in the RNA; for each G, a C; and for each T, an A. However, each A on the DNA guides the insertion of the pyrimidine uracil (U, from uridine triphosphate, UTP). There is no T in RNA.

***Quality control:*** Occasionally RNA polymerase will select and insert an incorrect, mismatched, ribonucleotide. When this occurs in bacteria (and perhaps in all organisms), the enzyme backs up, removes the incorrect nucleotide (and the one preceding it) and tries again. (Described by Zenkin *et al.*, in the 28 July 2006 issue of Science.)

- Synthesis of the RNA proceeds in the 52 '! 32 direction.
- As each nucleoside triphosphate is brought in to add to the 32 end of the growing strand, the two terminal phosphates are removed.
- When transcription is complete, the transcript is released from the polymerase and, shortly thereafter, the polymerase is released from the DNA.

Note that at any place in a DNA molecule, either strand may be serving as the template; that is, some genes "run" one way, some the other (and in a few remarkable cases, the same segment of double helix contains genetic information on both strands!). In all cases, however, RNA polymerase transcribes the DNA strand in its 32 '! 52 direction. A report in the 4 January 2001 issue of Nature shows that RNA polymerase actually tracks around the double helix of DNA. In vitro, at least, when RNA polymerase is immobilized, it spins the DNA molecule around and around as it moves along the molecule. Whether it is the polymerase or the DNA that does the spinning in vivo remains to be determined.

## Types of RNA

Sedimentation pattern produced by high-speed centrifugation of RNA extracted from the precursors of rabbit red blood cells. The discrete bands represent particular classes of RNA.

The transfer RNAs band at about 4S. The ribosomal RNAs of eukaryotes sediment at 5S, 5.8S, 18S, and 28S. (The larger the sedimentation unit, S, the larger the molecule — but not proportionally.) The RNA forming the band at 9S is the messenger RNA for the synthesis of hemoglobin, the major protein synthesized in these cells. In most types of cells, the messenger RNAs are extremely heterogenous, with small amounts distributed from 6S to 25S.

Several types of RNA are synthesized in the nucleus of eukaryotic cells. Of particular interest are:

- messenger RNA (mRNA). This will later be translated into a polypeptide.
- ribosomal RNA (rRNA). This will be used in the building of ribosomes: machinery for synthesizing proteins by translating mRNA.
- transfer RNA (tRNA). RNA molecules that carry amino acids to the growing polypeptide.
- small nuclear RNA (snRNA). DNA transcription of the genes for mRNA, rRNA, and tRNA produces large precursor molecules ("primary transcripts") that must be processed within the nucleus to produce the functional molecules for export to the cytosol. Some of these processing steps are mediated by snRNAs.
- small nucleolar RNA (snoRNA). These RNAs within the nucleolus have several functions (described below).

- microRNA (miRNA). These are tiny (~22 nucleotides) RNA molecules that appear to regulate the expression of messenger RNA (mRNA) molecules. [Discussion]
- XIST RNA. This inactivates one of the two X chromosomes in female vertebrates. [Discussion].

### *Messenger RNA (mRNA)*

Messenger RNA comes in a wide range of sizes reflecting the size of the polypeptide it encodes. Most cells produce small amounts of thousands of different mRNA molecules, each to be translated into a peptide needed by the cell. Many mRNAs are common to most cells, encoding "housekeeping" proteins needed by all cells (e.g. the enzymes of glycolysis). Other mRNAs are specific for only certain types of cells. These encode proteins needed for the function of that particular cell (e.g., the mRNA for hemoglobin in the precursors of red blood cells).

### *Ribosomal RNA (rRNA)*

There are 4 kinds. In eukaryotes, these are:

- 18S rRNA. One of these molecules, along with some 30 different protein molecules, is used to make the small subunit of the ribosome.
- 28S, 5.8S, and 5S rRNA. One each of these molecules, along with some 45 different proteins, are used to make the large subunit of the ribosome.

The S number given each type of rRNA reflects the rate at which the molecules sediment in the ultracentrifuge. The larger the number, the larger the molecule (but not proportionally).

The 28S, 18S, and 5.8S molecules are produced by the processing of a single primary transcript from a cluster of identical copies of a single gene. The 5S molecules are produced from a different cluster of identical genes.

### *Transfer RNA (tRNA)*

There are some 32 different kinds of tRNA in a typical eukaryotic cell.

- Each is the product of a separate gene.
- They are small (~4S), containing 73-93 nucleotides.

- Many of the bases in the chain pair with each other forming sections of double helix.
- The unpaired regions form 3 loops.
- Each kind of tRNA carries (at its 32 end) one of the 20 amino acids (thus most amino acids have more than one tRNA responsible for them).
- At one loop, 3 unpaired bases form an anticodon.
- Base pairing between the anticodon and the complementary codon on a mRNA molecule brings the correct amino acid into the growing polypeptide chain.

### Small Nuclear RNA (snRNA)

Approximately a dozen different genes for snRNAs, each present in multiple copies, have been identified. The snRNAs have various roles in the processing of the other classes of RNA.

For example, several snRNAs are part of the spliceosomes that participate in converting pre-mRNA into mRNA by excising the introns and splicing the exons.

### Small Nucleolar RNA (snoRNA)

As the name suggests, these small (60–300 nucleotides) RNAs are found in the nucleolus where they are responsible for several functions:

- Some participate in making ribosomes by helping to cut up the large RNA precursor of the 28S, 18S, and 5.8S molecules.
- Others chemically modify many of the nucleotides in rRNA, tRNA, and snRNA molecules, e.g., by adding methyl groups to ribose.
- Some have been implicated in the alternative splicing of pre-mRNA to different forms of mature mRNA.
- One snoRNA serves as the template for the synthesis of telomeres.

In vertebrates, the snoRNAs are made from introns removed during RNA processing.

### Noncoding RNA

Only messenger RNA encodes polypeptides. All the other classes of RNA, including types not mentioned here, are thus called noncoding RNA. Much remains to be learned about the function(s) of some of them. But, taken together, noncoding RNAs probably account for two-thirds of the transcription going on in the nucleus.

### *The RNA Polymerases*

The RNA polymerases are huge multi-subunit protein complexes. Three kinds are found in eukaryotes.

- RNA polymerase I (Pol I). It transcribes the rRNA genes for the precursor of the 28S, 18S, and 5.8S molecules (and is the busiest of the RNA polymerases).
- RNA polymerase II (Pol II; also known as RNAP II). It transcribes protein-encoding genes into mRNA (and also the snRNA genes).
- RNA polymerase III (Pol III). It transcribes the 5S rRNA genes and all the tRNA genes.

Two remarkable reports in the 8 June 2001 issue of Science show the structure of Pol II and reveal many details of how it uses the antisense strand of DNA to synthesize a strand of mRNA.

## RNA Processing: Pre-mRNA '! mRNA

All the primary transcripts produced in the nucleus must undergo processing steps to produce functional RNA molecules for export to the cytosol. We shall confine ourselves to a view of the steps as they occur in the processing of pre-mRNA to mRNA. Most eukaryotic genes are split into segments. In decoding the open reading frame of a gene for a known protein, one usually encounters periodic stretches of DNA calling for amino acids that do not occur in the actual protein product of that gene. Such stretches of DNA, which get transcribed into RNA but not translated into protein, are called introns. Those stretches of DNA that do code for amino acids in the protein are called exons. Examples:

- The gene for one type of collagen found in chickens is split into 52 separate exons.
- The gene for dystrophin, which is mutated in boys with muscular dystrophy, has 79 exons.
- Even the genes for rRNA and tRNA are split by introns.

- The human genome is estimated to contain some 180,000 exons. With a current estimate of 21,000 genes, the average exon content of our genes is about 9.

In general, introns tend to be much longer than exons. An average eukaryotic exon is only 140 nucleotides long, but one human intron stretches for 480,000 nucleotides!

Removal of the introns — and splicing the exons together — are among the essential steps in synthesizing mRNA.

### *The Steps of RNA Processing*

- Synthesis of the cap. This is a modified guanine (G) which is attached to the 52 end of the pre-mRNA as it emerges from RNA polymerase II (RNAP II). The cap
  - protects the RNA from being degraded by enzymes that degrade RNA from the 52 end;
  - serves as an assembly point for the proteins needed to recruit the small subunit of the ribosome to begin translation.
- Step-by-step removal of introns present in the pre-mRNA and splicing of the remaining exons. This step takes place as the pre-mRNA continues to emerge from RNAP II.
- Synthesis of the poly(A) tail. This is a stretch of adenine (A) nucleotides. When a special poly(A) attachment site in the pre-mRNA emerges from RNAP II, the transcript is cut there, and the poly(A) tail is attached to the exposed 32 end. This completes the mRNA molecule, which is now ready for export to the cytosol. (The remainder of the transcript is degraded, and the RNA polymerase leaves the DNA).

The cutting and splicing of mRNA must be done with great precision. If even one nucleotide is left over from an intron or one is removed from an exon, the reading frame from that point on will be shifted, producing new codons specifying a totally different sequence of amino acids from that point to the end of the molecule (which often ends prematurely anyway when the shifted reading frame generates a STOP codon).

The removal of introns and splicing of exons is done by spliceosomes. These are a complexes of 5 snRNA molecules and some 145 different proteins. The introns in most pre-mRNAs begin with a GU and end with an AG. Presumably these short sequences assist in guiding the spliceosome.

### *Visual Evidence*

The upper image is an electron micrograph of a mRNA-DNA hybrid molecule formed by mixing the messenger RNA (mRNA) from a clone of antibody-secreting cells with single-stranded DNA from the same kind of cells. The bar represents the length of 1000 bases.

The lower diagram is an interpretation of the micrograph. The solid line represents the DNA; the dotted line the mRNA. The loops ($I_A$, $I_B$, etc.) represent the introns that separate the exons encoding the domains of an antibody heavy chain:

- V = variable region
- $E_1$ = first constant region ($C_H1$) domain
- $E_H$ = hinge region
- $E_2$ and $E_3$ = the nucleotides encoding the two C-terminal domains ($C_H2$ and $C_H3$).

The unhybridized portion of the mRNA is its poly(A) tail.

### *Alternative Splicing*

The processing of pre-mRNA for many proteins proceeds along various paths in different cells or under different conditions. For example, early in the differentiation of a B cell (a lymphocyte that synthesizes an antibody) the cell first uses an exon that encodes a transmembrane domain that causes the molecule to be retained at the cell surface. Later, the B cell switches to using a different exon whose domain enables the protein to be secreted from the cell as a circulating antibody molecule.

Alternative splicing provides a mechanism for producing a wide variety of proteins from a small number of genes. While we humans may turn out to have only some 20 thousand genes, we probably make at least 10 times that number of different proteins.

It is now estimated that 92–94% of our genes produce pre-mRNAs that are alternatively-spliced. There is evidence that the pattern of alternative splicing differs consistently in different tissues and so must be regulated. But whether all the products are functional or that many are simply the outcome of an error-prone process remains to be seen.

Alternative splicing not only provides different proteins from a single gene but also different 3' UTRs and 5' UTRs. Although not translated into protein, these untranslated regions contain signals that, for example, dictate where in the cell that protein will accumulate. Two examples:

- The 3' UTR of the *bicoid* gene in Drosophila directs the mRNA to the anterior of the embryo;
- the same region in the *VegT* gene of Xenopus directs its mRNA to the vegetal pole of the embryo.

One of the most dramatic examples of alternative splicing is the *Dscam* gene in Drosophila. This single gene contains some 116 exons of which 17 are retained in the final mRNA. Some exons are always included; others are selected from an array. Theoretically this system is able to produce 38,016 different proteins. And, in fact, over 18,000 different ones have been found in Drosophila hemolymph.

These Dscam proteins are used to establish a unique identity for each neuron. It works like this. Each developing neuron draws upon the pool of thousands of possible different mRNAs to synthesize a dozen or so of them. Which ones are selected appears to be simply a matter of chance, but because of the size of the pool, each neuron will most likely end up with a unique set of a dozen or so Dscam proteins. As each developing neuron in the central nervous system sprouts dendrites and an axon, these express its unique collection of Dscam proteins. If the extensions of a single neuron should meet each other in the tangled web that is the hallmark of nervous tissue, they are repelled. In this way, thousands of different neurons can coexist in intimate contact without the danger of nonfunctional contacts between the various extensions of the same neuron.

We also have a *DSCAM* gene which is located on chromosome 21. When present in 3 copies, it is probably responsible for some of the features of Down syndrome (and accounts for its full name: Down Syndrome Cell Adhesion Molecule). Perhaps the incredible diversity of synaptic junctions in the mammalian c.n.s. ($\sim 10^{14}$) is mediated by alternative splicing of a limited number of gene transcripts. So, whether a particular segment of RNA will be retained as an exon or excised as an intron can vary under different circumstances, such as;

- what type of cell the gene is in;
- what stage of differentiation that cell is passing through;
- what extracellular signals that cell is receiving.

Clearly the switching to an alternate splicing pathway must be closely regulated.

### *Trans-splicing*

Most genes are transcribed and their transcripts processed as described above. RNA polymerase travels down a single strand of a singe gene locus to form pre-mRNA that is processed (including removal of introns) to form the mature mRNA. But there are exceptions. A number of cases have been found where two different precursor transcripts have been spliced together to form the final RNA molecule. The phenomenon is called *trans*-splicing. Examples: synthesis of a single RNA molecule by splicing together transcripts from ;

- loci located far apart on the same chromosome;
- opposite strands of the same gene locus;
- the two alleles of the gene on their separate (homologous) chromosomes.

The biological importance of these *trans*-spliced transcripts is still unknown for most of them.

### *Why Split Genes?*

Perhaps during evolution, eukaryotic genes have been assembled from smaller, primitive genes - today's exons. Some proteins, like the antibodies mentioned in the previous section, are organized in a set of separate sections or domains each with a special function to perform in the complete molecule. Each domain is encoded by a separate exon. Having the different functional parts of the antibody molecule encoded by separate exons makes it possible to use these units in different combinations. Thus a set of exons in the genome may be the genetic equivalent of the various modular pieces in a box of "Lego" for children to assemble in whatever forms they wish.

But the boundaries of other exons do not seem to correspond to domain boundaries of the protein. So perhaps the major benefit of split genes is simply the opportunity they provide for making many different proteins from a single gene through alternative splicing.

### *Summary*

Gene expression occurs in two steps:

- transcription of the information encoded in DNA into a molecule of RNA (described here) and
- translation of the information encoded in the nucleotides of mRNA into a defined sequence of amino acids in a protein (discussed in Gene Translation: RNA → Protein).

## Messenger RNA

Messenger RNA (mRNA) is a molecule of RNA encoding a chemical "blueprint" for a protein product. mRNA is transcribed from a DNA template, and carries coding information to the sites of protein synthesis: the ribosomes. Here, the nucleic acid polymer is translated into a polymer of amino acids: a protein. In mRNA as in DNA, genetic information is encoded in the sequence of nucleotides arranged into codons consisting of three bases each. Each codon encodes for a specific amino acid, except the stop codons that terminate protein synthesis. This process requires two other types of RNA: transfer RNA (tRNA) mediates recognition of the codon and provides the corresponding amino acid, while ribosomal RNA (rRNA) is the central component of the ribosome's protein manufacturing machinery.

### *Synthesis, Processing, and Function*

The brief existence of an mRNA molecule begins with transcription and ultimately ends in degradation. During its life, an mRNA molecule may also be processed, edited, and transported prior to translation. Eukaryotic mRNA molecules often require extensive processing and transport, while prokaryotic molecules do not.

### *Transcription*

During transcription, RNA polymerase makes a copy of a gene from the DNA to mRNA as needed. This process is similar in eukaryotes and prokaryotes. One notable difference, however, is that prokaryotic RNA polymerase associates with mRNA processing enzymes during transcription so that processing can proceed quickly after the start of transcription. The short-lived, unprocessed or partially processed, product is termed *pre-mRNA*; once completely processed, it is termed *mature mRNA*.

### *Eukaryotic Pre-mRNA Processing*

Processing of mRNA differs greatly among eukaryotes, bacteria and archea. Non-eukaryotic mRNA is essentially mature upon transcription and requires no processing, except in rare cases. Eukaryotic pre-mRNA, however, requires extensive processing.

### *5' Cap Addition*

A *5' cap* (also termed an RNA cap, an RNA 7-methylguanosine cap or an RNA $m^7G$ cap) is a modified guanine nucleotide that has been added to the "front" or 5' end of a eukaryotic messenger RNA shortly after the start of transcription. The 5' cap consists of a terminal 7-methylguanosine residue which is linked through a 5'-5'-triphosphate bond to the first transcribed nucleotide.

Its presence is critical for recognition by the ribosome and protection from RNases. Cap addition is coupled to transcription, and occurs co-transcriptionally, such that each influences the other. Shortly after the start of transcription, the 5' end of the mRNA being synthesized is bound by a cap-synthesizing complex associated with RNA polymerase. This enzymatic complex catalyzes the chemical reactions that are required for mRNA capping. Synthesis proceeds as a multi-step biochemical reaction.

### *Splicing*

Splicing is the process by which pre-mRNA is modified to remove certain stretches of non-coding sequences called introns; the stretches that remain include protein-coding sequences and are called exons. Sometimes pre-mRNA messages may be spliced in several different ways, allowing a single gene to encode multiple proteins. This process is called alternative splicing. Splicing is usually performed by an RNA-protein complex called the spliceosome, but some RNA molecules are also capable of catalyzing their own splicing.

## Ribozyme

A ribozyme (from ribonucleic acid enzyme, also called RNA enzyme or catalytic RNA) is an RNA molecule possessing a well defined tertiary structure that enables it to catalyze a chemical reaction. Many natural ribozymes catalyze either the hydrolysis of one of their own phosphodiester bonds (self-cleaving ribozymes), or the hydrolysis of bonds in other RNAs, but they have also been found to catalyze the aminotransferase activity of the ribosome. Investigators studying the origin of life have produced ribozymes in the laboratory that are capable of catalyzing their own synthesis under very specific conditions, such as an RNA polymerase ribozyme. Mutagenesis and selection has been performed resulting in isolation of improved variants of the "Round-18" polymerase ribozyme from 2001. "B6.61" is able to add up to 20 nucleotides to a primer template in 24 hours, until it decomposes by hydrolysis of its phosphodiester bonds.

Some ribozymes may play an important role as therapeutic agents, as enzymes which tailor defined RNA sequences, as biosensors, and for applications in functional genomics and gene discovery.

### *Discovery*

Before the discovery of ribozymes, enzymes, which are defined as catalytic proteins, were the only known biological catalysts. In 1967, Carl Woese, Francis Crick, and Leslie Orgel were the first to suggest that RNA could act as a catalyst. This idea was based upon the discovery that RNA can form complex secondary structures. The first ribozymes were discovered in the 1980s by Thomas R. Cech, who was studying RNA splicing in the ciliated protozoan *Tetrahymena thermophila* and Sidney Altman, who was working on the bacterial RNase P complex.

These ribozymes were found in the intron of an RNA transcript, which removed itself from the transcript, as well as in the RNA component of the RNase P complex, which is involved in the maturation of pre-tRNAs. In 1989, Thomas R. Cech and Sidney Altman won the Nobel Prize in chemistry for their "discovery of catalytic properties of RNA." The term *ribozyme* was first introduced by Kelly Kruger *et al.* in 1982 in a paper published in *Cell.*

It had been a firmly established belief in biology that catalysis was reserved for proteins. In retrospect, catalytic RNA makes a lot of sense. This is based on the old question regarding the origin of life: Which comes first, enzymes that do the work of the cell or nucleic acids that carry the information required to produce the enzymes? Nucleic acids as catalysts circumvents this problem.

In the 1970s Thomas Cech, at the University of Colorado at Boulder, was studying the excision of introns in a ribosomal RNA gene in *Tetrahymena thermophila*. While trying to purify the enzyme responsible for splicing reaction, he found that intron could be spliced out in the absence of any added cell extract. As much as they tried, Cech and his colleagues could not identify any protein associated with the splicing reaction. After much work, Cech proposed that the intron sequence portion of the RNA could break and reform phosphodiester bonds. At about the same time, Sidney Altman, who is a Professor at Yale University, was studying the way tRNA molecules are processed in the cell when he and his colleagues isolated an enzyme called RNase-P, which is responsible for conversion of a precursor tRNA into the active tRNA.

Much to their surprise, they found that RNase-P contained RNA in addition to protein and that RNA was an essential component of the active enzyme. This was such a foreign idea that they had difficulty publishing their findings. The following year, Altman demonstrated that RNA can act as a catalyst by showing that the RNase-P RNA subunit could catalyze the cleavage of precursor tRNA into active tRNA in the absence of any protein component. Since Cech's and Altman's discovery, other investigators have discovered other examples of self-cleaving RNA or catalytic RNA molecules. Many ribozymes have either a hairpin – or hammerhead – shaped active center and a unique secondary structure that allows them to cleave other RNA molecules at specific sequences. It is now possible to make ribozymes that will specifically cleave any RNA molecule. These RNA catalysts may have pharmaceutical applications. For example, a ribozyme has been designed to cleave the RNA of HIV. If such a ribozyme was made by a cell, all incoming virus particles would have their RNA genome cleaved by the ribozyme, which would prevent infection.

### *Activity*

Although most ribozymes are quite rare in the cell, their roles are sometimes essential to life. For example, the functional part of the ribosome, the molecular machine that translates RNA into proteins, is fundamentally a ribozyme, composed of RNA tertiary structural motifs that are often coordinated to metal ions such as $Mg^{2+}$ as cofactors. There is no requirement for divalent cations in a five-nucleotide RNA that can catalyze *trans*-phenylalanation of a four-nucleotide substrate which has three base complementary sequence with the catalyst. The catalyst and substrate were devised by truncation of the C3 ribozyme.

RNA can also act as a hereditary molecule, which encouraged Walter Gilbert to propose that in the distant past, the cell used RNA as both the genetic material and the structural and catalytic molecule, rather than dividing these functions between DNA and protein as they are today. This hypothesis became known as the "RNA world hypothesis" of the origin of life.

If ribozymes were the first molecular machines used by early life, then today's remaining ribozymes — such as the ribosome machinery — could be considered living fossils of a life based primarily on nucleic acids. A recent test-tube study of prion folding suggests that an RNA may catalyze the pathological protein conformation in the manner of a chaperone enzyme.

### *Known Ribozymes*

Naturally occurring ribozymes include:

- Peptidyl transferase 23S rRNA
- RNase P
- Group I and Group II introns
- GIR1 branching ribozyme
- Leadzyme - Although initially created *in vitro*, natural examples have been found
- Hairpin ribozyme
- Hammerhead ribozyme
- HDV ribozyme

- Mammalian CPEB3 ribozyme
- VS ribozyme
- *glmS* ribozyme
- CoTC ribozyme.

### *Artificial Ribozymes*

Since the discovery of ribozymes that exist in living organisms, there has been interest in the study of new synthetic ribozymes made in the laboratory. For example, artificially-produced self-cleaving RNAs that have good enzymatic activity have been produced.

Tang and Breaker isolated self-cleaving RNAs by in vitro selection of RNAs originating from random-sequence RNAs. Some of the synthetic ribozymes that were produced had novel structures, while some were similar to the naturally occurring hammerhead ribozyme. The techniques used to discover artificial ribozymes involve Darwinian evolution. This approach takes advantage of RNA's dual nature as both a catalyst and an informational polymer, making it easy for an investigator to produce vast populations of RNA catalysts using polymerase enzymes. The ribozymes are mutated by reverse transcribing them with reverse transcriptase into various cDNA and amplified with mutagenic PCR. The selection parameters in these experiments often differ. One approach for selecting a ligase ribozyme involves using biotin tags, which are covalently linked to the substrate. If a molecule possesses the desired ligase activity, a streptavidin matrix can be used to recover the active molecules.

Lincoln and Joyce developed an RNA enzyme system capable of self replication in about an hour. By utilizing molecular competition (*in vitro* evolution) of a candidate enzyme mixture, a pair of RNA enzymes emerged, in which each synthesizes the other from synthetic oligonucleotides, with no protein present.

### *Applications*

A type of synthetic ribozyme directed against HIV RNA called gene shears has been developed and has entered clinical testing for HIV infection.

## Editing

In some instances, an mRNA will be edited, changing the nucleotide composition of that mRNA. An example in humans is the apolipoprotein B mRNA, which is edited in some tissues, but not others. The editing creates an early stop codon, which upon translation, produces a shorter protein.

## Polyadenylation

Polyadenylation is the addition of a poly(A) tail to an RNA molecule. The poly(A) tail consists of multiple adenosine monophosphates; in other words, it is a stretch of RNA which only has adenine bases. In eukaryotes, polyadenylation is part of the process that produces mature messenger RNA (mRNA) for translation. It therefore forms part of the larger process of gene expression.

The process of polyadenylation begins as the transcription of a gene finishes. The 3'-most segment of the newly-made RNA is first cleaved off by a set of proteins; these proteins then synthesize the poly(A) tail at the RNA's 3' end. In some genes these proteins may add a poly(A) tail at any one of several possible sites, polyadenylation can therefore produce more than one transcript from a single gene, similar to alternative splicing.

The poly(A) tail is important for the nuclear export, translation and stability of mRNA. The tail is shortened over time and when it is short enough, the mRNA is enzymatically degraded. However, in a few cell types, mRNAs with short poly(A) tails are stored for later activation by re-polyadenylation in the cytosol.

In contrast, when polyadenylation occurs in bacteria, it promotes RNA degradation. This is also sometimes the case for eukaryotic non-coding RNAs. The wide distribution of polyadenylation among living organisms indicates that this process evolved early in the history of life on Earth.

## Primer on RNA

RNAs are a type of large biological molecules, whose individual building blocks are called nucleotides. The name *poly(A) tail* (for polyadenylic acid tail) reflects the way RNA nucleotides are abbreviated, with a letter for the base the nucleotide contains (A for adenine, C for cytosine, G for guanine and U for uracil). RNAs are produced (*transcribed*) from a DNA template. By convention, RNA sequences are written in a 5' to 3' direction. The 5' end is the part of the RNA molecule that is transcribed first, and the 3' end is transcribed last. The 3' end is also where the poly(A) tail is found on polyadenylated RNAs. Messenger RNA (mRNA) is RNA that has a coding region that acts as a template for protein synthesis (*translation*). The rest of the mRNA, the *untranslated regions*, tune how active the mRNA is. There are also many RNAs that are not translated, called non-coding RNAs. Like the untranslated regions, many of these non-coding RNAs have regulatory roles.

## Nuclear Polyadenylation

### *Function*

In nuclear polyadenylation, a poly(A) tail is added to an RNA at the end of transcription. On mRNAs, the poly(A) tail protects the mRNA molecule from enzymatic degradation in the cytoplasm and aids in transcription termination, export of the mRNA from the nucleus, and translation. Almost all eukaryotic mRNAs are polyadenylated, with the exception of animal replication-dependent histone mRNAs. These are the only mRNAs in eukaryotes that lack a poly(A) tail, ending instead in a stem-loop structure followed by a purine-rich sequence, termed histone downstream element, that directs where the RNA is cut so that the 3' end of the histone mRNA is formed.

Many eukaryotic non-coding RNAs are always polyadenylated at the end of transcription. There are small RNAs where the poly(A) tail is only seen in intermediary forms and not in the mature RNA as the ends are removed during processing, notably microRNAs. But for many long noncoding RNAs – a seemingly large group of regulatory RNAs that for example includes the RNA Xist which mediates X chromosome inactivation – a poly(A) tail is part of the mature RNA.

### *Mechanism*

The polyadenylation machinery in the nucleus of eukaryotes works on products of RNA polymerase II, such as precursor mRNA. Here, a multi-protein complex cleaves the 3'-most part of a newly produced RNA and polyadenylates the end produced by this cleavage. The cleavage is catalysed by the enzyme CPSF and occurs 10–30 nucleotides downstream of its binding site.

This site is often the sequence AAUAAA on the RNA, but variants of it exist that bind more weakly to CPSF. Two other proteins add specificity to the binding to an RNA: CstF and CFI. CstF binds to a GU-rich region further downstream of CPSF's site. CFI recognises a third site on the RNA (a set of UGUAA sequences in mammals) and can recruit CPSF even if the AAUAAA sequence is missing. The polyadenylation signal – the sequence motif recognised by the RNA cleavage complex – varies between groups of eukaryotes. Most human polyadenylation sites contain the AAUAAA sequence, but this sequence is less common in plants and fungi.

The RNA is cleaved right after transcription, as CstF also binds to RNA polymerase II. Cleavage also involves the protein CFII, though it is unknown how. The cleavage site associated with a polyadenylation signal can vary up to some 50 nucleotides. When the RNA is cleaved, polyadenylation starts, catalysed by polyadenylate polymerase.

Polyadenylate polymerase builds the poly(A) tail by adding adenosine monophosphate units from adenosine triphosphate to the RNA, cleaving off pyrophosphate. Another protein, PAB2, binds to the new, short poly(A) tail and increases the affinity of polyadenylate polymerase for the RNA. When the poly(A) tail is approximately 250 nucleotides long the enzyme can no longer bind to CPSF and polyadenylation stops, thus determining the length of the poly(A) tail. CPSF is in contact with RNA polymerase II, allowing it to tell the polymerase to terminate transcription. The polyadenylation machinery is also physically linked to the spliceosome, a complex that removes introns from RNAs.

### *Downstream Effects*

The poly(A) tail acts as the binding site for poly(A)-binding protein. Poly(A)-binding protein promotes export from the nucleus and translation, and inhibits degradation. This protein binds to the poly(A) tail prior to mRNA export from the nucleus and in yeast also recruits poly(A) nuclease, an enzyme that shortens the poly(A) tail and allows the export of the mRNA. Poly(A)-binding protein is exported to the cytoplasm with the

RNA. mRNAs that are not exported are degraded by the exosome. Poly(A)-binding protein also can bind to, and thus recruit, several proteins that affect translation, one of these is initiation factor-4G which in turn recruits the 40S ribosomal subunit. However, a poly(A) tail is not required for the translation of all mRNAs.

### Deadenylation

In eukaryotic somatic cells, the poly(A) tail of most mRNAs in the cytoplasm gradually get shorter, and mRNAs with shorter poly(A) tail are translated less and degraded sooner. However, it can take many hours before an mRNA is degraded. This deadenylation and degradation process can be accelerated by microRNAs complementary to the 3' untranslated region of an mRNA. In immature egg cells, mRNAs with shortened poly(A) tails are not degraded, but are instead stored without being translated. They are then activated by cytoplasmic polyadenylation after fertilisation, during egg activation. In animals, poly(A) ribonuclease (PARN) can bind to the 5' cap and remove nucleotides from the poly(A) tail. The level of access to the 5' cap and poly(A) tail is important in controlling how soon the mRNA is degraded. PARN deadenylates less if the RNA is bound by the initiation factors 4E (at the 5' cap) and 4G (at the poly(A) tail), so this is why translation reduces deadenylation. The rate of deadenylation may also be regulated by RNA-binding proteins. Once the poly(A) tail is removed, the decapping complex removes the 5' cap, leading to a degradation of the RNA. Several other enzymes that seem to be involved in deadenylation have been identified in yeast.

### Alternative Polyadenylation

Many protein-coding genes have more than one polyadenylation site, so a gene can code for several mRNAs that differ in their 3' end. Since alternative polyadenylation changes the length of the 3' untranslated region, it can change which binding sites for microRNAs the 3' untranslated region contains. MicroRNAs tend to repress translation and promote degradation of the mRNAs they bind to, although there are examples of microRNAs that stabilise transcripts. Alternative polyadenylation can also shorten the coding region, thus making the mRNA code for a different protein, but this is much less common than just shortening the 3' untranslated region. The choice of poly(A) site can be influenced by extracellular stimuli and depends on the expression of the proteins that take part in polyadenylation. For example, the expression of CstF-64, a subunit of cleavage stimulatory factor (CstF), increases in macrophages in response to lipopolysaccharides (a group of bacterial compounds that trigger an immune response). This results in the selection of weak poly(A) sites and thus shorter transcripts.

This removes regulatory elements in the 3' untranslated regions of mRNAs for defense-related products like lysozyme and TNF-α. These mRNAs then have longer half-lives and produce more of these proteins. RNA-binding proteins other than those in the polyadenylation machinery can also affect whether a polyadenyation site is used,as can DNA methylation near the polyadenylation signal.

## Cytoplasmic Polyadenylation

There is polyadenylation in the cytosol of some animal cell types, namely in the germ line, during early embryogenesis and in post-synaptic sites of nerve cells. This lengthens the poly(A) tail of an mRNA with a shortened poly(A) tail, so that the mRNA will be translated. These shortened poly(A) tails are often less than 20 nucleotides, and are lengthened to around 80–150 nucleotides.

In the early mouse embryo, cytoplasmic polyadenylation of maternal RNAs from the egg cell allows the cell to survive and grow even though transcription does not start until the middle of the 2-cell stage (4-cell stage in human). In the brain, cytoplasmic polyadenylation is active during learning and could play a role in long-term potentiation, which is the strengthening of the signal transmission from a nerve cell to another in response to nerve impulses and is important for learning and memory formation.

Cytoplasmic polyadenylation requires the RNA-binding proteins CPSF and CPEB, and can involve other RNA-binding proteins like Pumilio. Depending on the cell type, the polymerase can be the same type of polyadenylate polymerase (PAP) that is used in the nuclear process, or the cytoplasmic polymerase GLD-2.

### *Tagging for Degradation in Eukaryotes*

For many non-coding RNAs, including tRNA, rRNA, snRNA and snoRNA, polyadenylation is a way of marking the RNA for degradation, in at least yeast.

This polyadenylation is done in the nucleus by the TRAMP complex, which adds a tail that is around 40 nucleotides long to the 3' end. The RNA is then degraded by the exosome. Poly(A) tails have also been found on human rRNA fragments, both the form of homopolymeric (A only) and heterpolymeric (mostly A) tails.

### *In Prokaryotes and Organelles*

In many bacteria, both mRNAs and non-coding RNAs can be polyadenylated. This poly(A) tail promotes degradation by the degradosome, which contains two RNA-degrading enzymes: polynucleotide phosphorylase and RNase E. Polynucleotide phosphorylase binds to the 3' end of RNAs and the 3' extension provided by the poly(A) tail allows it to bind to the RNAs whose secondary structure would otherwise block the 3' end. Successive rounds of polyadenylation and degradation of the 3' end by polynucleotide phosphorylase allows the degradosome to overcome these secondary structures.

The poly(A) tail can also recruit RNases that cut the RNA in two. These bacterial poly(A) tails are about 30 nucleotides long. In as different groups as animals and trypanosomes, the mitochondria contain both stabilising and destabilising poly(A) tails.

Destabilising polyadenylation targets both mRNA and noncoding RNAs. The poly(A) tails are 43 nucleotides long on average. The stabilising ones start at the stop codon, and without them the stop codon (UAA) is not complete as the genome only encodes the U or UA part. Plant mitochondria only have destabilising polyadenylation, and yeast mitochondria have no polyadenylation at all.

While many bacteria and mitochondria have polyadenylate polymerases, they also have another type of polyadenylation, performed by polynucleotide phosphorylase itself.

This enzyme is found in bacteria, mitochondria, plastids and as a constituent of the archeal exosome (in those archaea that have an exosome). It can synthesise a 3' extension where the vast majority of the bases are adenines. Like in bacteria, polyadenylation by polynucleotide phosphorylase promotes degradation of the RNA in plastids and likely also archaea.

### *Evolution*

Although polyadenylation is seen in almost all organisms, it is not universal. However, the wide distribution of this modification and the fact that it is present in organisms from all three domains of life implies that the last universal common ancestor of all living organisms probably had some form of polyadenylation system. A few organisms do not polyadenylate mRNA, which implies that they have lost their polyadenylation machineries during evolution. Although no examples of eukaryotes that lack polyadenylation are known, mRNAs from the bacterium *Mycoplasma gallisepticum* and the salt-tolerant archaean *Haloferax volcanii* lack this modification.

The most ancient polyadenylating enzyme is polynucleotide phosphorylase. This enzyme is part of both the bacterial degradosome and archaeal exosome, two closely related complexes that recycle RNA into nucleotides. This enzyme degrades RNA by attacking the bond between the 3'-most nucleotides with a phosphate, breaking off a diphosphate nucleotide.

This reaction is reversible, and so the enzyme can also extend RNA with more nucleotides. The heteropolymeric tail added by polynucleotide phosphorylase is very rich in adenine. The choice of adenine is most likely the result of higher ADP concentrations than other nucleotides as a result of using ATP as an energy currency, making it more likely to be incorporated in this tail in early lifeforms. It has been suggested that the involvement of adenine-rich tails in RNA degradation prompted the later evolution of polyadenylate polymerases (the enzymes that produce poly(A) tails with no other nucleotides in them).

Polyadenylate polymerases are not as ancient. They have separately evolved in both bacteria and eukaryotes from CCA-adding enzyme, which is the enzyme that completes the 3' ends of tRNAs. Its catalytic domain is homologous to that of other polymerases. Likely, the horizontal transfer of bacterial CCA-adding enzyme to eukaryotes allowed the archaeal-like CCA-adding enzyme to switch function to a poly(A) polymerase. Some lineages, like archaea and cyanobacteria, never evolved a polyadenylate polymerase.

### History

Polyadenylation was first identified in 1960 as an enzymatic activity in extracts made from cell nuclei that could polymerise ATP, but not ADP, into polyadenine. Although identified in many types of cells, this activity had no known function until 1971, when poly(A) sequences were found in mRNAs. Initially, the only function of these sequences was thought to be protection of the 3' end of the RNA from nucleases, but later the specific roles of polyadenylation in nuclear export and translation were identified. The polymerases responsible for polyadenylation were first purified and characterized in the 1960s and 1970s, but the large number of accessory proteins that control this process were only discovered in the early 1990s. Polyadenylation is the covalent linkage of a polyadenylyl moiety to a messenger RNA molecule. In eukaryotic organisms, most messenger RNA (mRNA) molecules are polyadenylated at the 3' end. The poly(A) tail and the protein bound to it aid in protecting mRNA from degradation by exonucleases. Polyadenylation is also important for transcription termination, export of the mRNA from the nucleus, and translation. mRNA can also be polyadenylated in prokaryotic organisms, where poly(A) tails act to facilitate, rather than impede, exonucleolytic degradation. Polyadenylation occurs during and immediately after transcription of DNA into RNA. After transcription has been terminated, the mRNA chain is cleaved through the action of an endonuclease complex associated with RNA polymerase. After the mRNA has been cleaved, around 250 adenosine residues are added to the free 3' end at the cleavage site. This reaction is catalyzed by polyadenylate polymerase. Just as in alternative splicing, there can be more than one polyadenylation variant of a mRNA.

### Transport

Another difference between eukaryotes and prokaryotes is mRNA transport. Because eukaryotic transcription and translation is compartmentally separated, eukaryotic mRNAs must be exported from the nucleus to the cytoplasm. Mature mRNAs are recognized by their processed modifications and then exported through the nuclear pore. In neurons mRNA must be transported from the soma to the dendrites where local translation occurs in response to external stimuli. Many messages are marked with so-called "zip codes" which targets their transport to a specific location.

### Translation

Because prokaryotic mRNA does not need to be processed or transported, translation by the ribosome can begin immediately after the end of transcription. Therefore, it can be said that prokaryotic translation is *coupled* to transcription and occurs *co-transcriptionally.*

Eukaryotic mRNA that has been processed and transported to the cytoplasm (i.e. mature mRNA) can then be translated by the ribosome. Translation may occur at ribosomes free-floating in the cytoplasm, or directed to the endoplasmic reticulum by the signal recognition particle. Therefore, unlike in prokaryotes, eukaryotic translation *is not* directly coupled to transcription.

### Structure

### 5' cap

The 5' cap is a specially altered nucleotide on the 5' end of precursor messenger RNA and some other primary RNA transcripts as found in eukaryotes. The process of 5' capping is vital to creating mature messenger RNA, which is then able to undergo translation. Capping ensures the messenger RNA's stability while it undergoes translation in the process of protein synthesis, and is a highly regulated process that occurs in the cell nucleus.

### 5' Cap Structure

The 5' cap is found on the 5' end of an mRNA molecule and consists of a guanine nucleotide connected to the mRNA via an unusual 5' to 5' triphosphate linkage. This guanosine is methylated on the 7 position directly after capping *in vivo* by a methyl transferase. It is referred to as a 7-methylguanosine cap, abbreviated $m^7G$.

Further modifications include the possible methylation of the 2' hydroxy-groups of the first 2 ribose sugars of the 5' end of the mRNA. The methylation of both 2' hydroxy-groups is shown on the diagram. The 5' cap looks like the 3' end of an RNA molecule (the 5' carbon of the cap ribose is bonded, and the 3' unbonded). This provides significant resistance to 5' exonucleases.

### Capping Process

The starting point is the unaltered 5' end of an RNA molecule. This features a final nucleotide followed by three phosphate groups attached to the 5' carbon.

1. One of the terminal phosphate groups is removed (by RNA terminal phosphatase), leaving two terminal phosphates.
2. GTP is added to the terminal phosphates (by a guanylyl transferase), losing two phosphate groups (from the GTP) in the process. This results in the 5' to 5' triphosphate linkage.
3. The 7-nitrogen of guanine is methylated (by a methyl transferase).
4. Other methyltransferases are optionally used to carry out methylation of 5' proximal nucleotides.

### 5' Capping Targeting

The Capping Enzyme Complex (CEC) required for capping is found bound to the RNA polymerase II before transcription starts. As soon as the 5' end of the new transcript emerges the enzymes transfer to it and begin the capping process (this is a similar kind of mechanism to ensure capping as for polyadenylation).

The enzymes for capping can only bind to RNA polymerase II ensuring specificity to only these transcripts, which are almost entirely mRNA.

### 5' Cap Function

The 5' cap has 4 main functions:

1. Regulation of nuclear export.
2. Prevention of degradation by exonucleases.
3. Promotion of translation.
4. Promotion of 5' proximal intron excision.

Nuclear export of RNA is regulated by the Cap binding complex (CBC), which binds exclusively to capped RNA. The CBC is then recognized by the nuclear pore complex and exported. Once in the cytoplasm after the pioneer round of translation, the CBC is replaced by the translation factors eIF-4E and eIF-4G. This complex is then recognized by other translation initiation machinery including the ribosome.

Cap prevents 5' degradation in two ways. First, degradation of the mRNA by 5' exonucleases is prevented (as mentioned above) by functionally looking like a 3' end. Second, the CBC complex and the eIF-4E/eIF-4G block the access of decapping enzymes to the cap. This increases the half-life of the mRNA, essential in eukaryotes as the export process takes significant time.

Decapping of an mRNA is catalyzed by the decapping complex made up of at least Dcp1 and Dcp2, which must compete with eIF-4E to bind the cap. Thus the 5' cap is a marker of an actively translating mRNA and is used by cells to regulate mRNA half-lives in response to new stimuli. Undesirable mRNAs are sent to P-bodies for temporary storage or decapping, the details of which are still being resolved.

The mechanism of 5' proximal intron excision promotion is not well understood, but the 5' cap appears to loop around and interact with the spliceosome in the splicing process, promoting intron excision.

## Messenger RNA Decapping

The process of messenger RNA decapping consists of hydrolysis of the 5' cap structure on the RNA exposing a 5' monophosphate. This 5' monophosphate is a substrate for the exonuclease Xrn1 and the message is quickly destroyed. There are many situations which may lead to the removal of the cap, some of which are discussed below.

### *Translation and Decay*

Inside cells there is a balance between the processes of translation and mRNA decay. Messages which are being actively translated are bound by polysomes and the initiation factors eIF-4E and eIF-4G. This blocks access to the cap by the decapping enzyme DCP2 and protects the message. In nutrient-starvation conditions or viral infection, translation may be compromised and decapping is stimulated. This balance is reflected in the size and abundance of the cytoplasmic structures known as P-bodies

### *Specific Decay Pathways*

A number of specific decay pathways exist that recognize aberrant messages and promote their decapping. Nonsense mediated decay recognizes premature stop codons and promotes decapping as well as decay by the exosome. Certain classes of miRNA have also been shown to stimulate decapping.

### *Coding Regions*

The coding region of a gene is that portion of a gene's DNA or RNA, composed of exons, that codes for protein. The region is bounded nearer the 5' end by a start codon and nearer the 3' end with a stop codon. The coding region in mRNA is bounded by the five prime untranslated region and the three prime untranslated region, which are also parts of the exons.

The coding region of an organism is sum total of the organism's genome that is composed of gene coding regions. Coding regions are composed of codons, which are decoded and translated (in eukaryotes usually into one and in prokaryotes usually into several) proteins by the ribosome. Coding regions begin with the start codon and end with a stop codon. Generally, the start codon is an AUG triplet and the stop codon is UAA, UAG, or UGA. The coding regions tend to be stabilised by internal base pairs, this impedes degradation. In addition to being protein-coding, portions of coding regions may serve as regulatory sequences in the pre-mRNA as exonic splicing enhancers or exonic splicing silencers.

## Open Reading Frame

In molecular genetics, an open reading frame (ORF) is a DNA sequence that does not contain a stop codon in a given reading frame.

### *Significance*

One common use of open reading frames are as one piece of evidence to assist in gene prediction. Long ORFs are often used, along with other evidence, to initially identify candidate protein coding regions in a DNA sequence. The presence of an ORF does not necessarily mean that the region is ever translated. For example in a randomly generated DNA sequence with an equal percentage of each nucleotide, a stop-codon would be expected once every 21 codons. A simple gene prediction algorithm for prokaryotes might look for a start codon followed by an open reading frame that is long enough to encode a typical protein, where the codon usage of that region matches the frequency characteristic for the given organism's coding regions. By itself even a long open reading frame is not conclusive evidence for the presence of a gene.

### *Example*

If a portion of a genome has been sequenced (e.g. 5'-ATCTAAAATGGGTGCC-3'), ORFs can be located by examining each of the three possible reading frames on each strand. In this sequence two out of three possible reading frames are entirely open, meaning that they do not contain a stop codon:

1. ...A TCT AAA ATG GGT GCC...
2. ...AT CTA AAA TGG GTG CC...
3. ...ATC TAA AAT GGG TGC C...

Possible stop codons in DNA are "TGA", "TAA" and "TAG". Thus, the last reading frame in this example contains a stop codon (*TAA*), unlike the first two.

### *Untranslated Regions*

Untranslated regions (UTRs) are sections of the mRNA before the start codon and after the stop codon that are not translated, termed the five prime untranslated region (5' UTR) and three prime untranslated region (3' UTR), respectively. These regions are transcribed with the coding region and thus are exonic as they are present in the mature mRNA. Several roles in gene expression have been attributed to the untranslated regions, including mRNA stability, mRNA localization, and translational efficiency. The ability of a UTR to perform these functions depends on the sequence of the UTR and can differ between mRNAs. The stability of mRNAs may be controlled by the 5' UTR and/or 3' UTR due to varying affinity for RNA degrading enzymes called ribonucleases and for ancillary proteins that can promote or inhibit RNA degradation.

Translational efficiency, including sometimes the complete inhibition of translation, can be controlled by UTRs. Proteins that bind to either the 3' or 5' UTR may affect translation by influencing the ribosome's ability to bind to the mRNA. MicroRNAs bound to the 3' UTR also may affect translational efficiency or mRNA stability.

Cytoplasmic localization of mRNA is thought to be a function of the 3' UTR. Proteins that are needed in a particular region of the cell can actually be translated there; in such a case, the 3' UTR may contain sequences that allow the transcript to be localized to this region for translation.

Some of the elements contained in untranslated regions form a characteristic secondary structure when transcribed into RNA. These structural mRNA elements are involved in regulating the mRNA. Some, such as the SECIS element, are targets for proteins to bind. One class of mRNA element, the riboswitches, directly bind small molecules, changing their fold to modify levels of transcription or translation. In these cases, the mRNA regulates itself.

### *5' UTR*

The five prime untranslated region (5' UTR), also known as the leader sequence, is a particular section of messenger RNA (mRNA) and the DNA that codes for it.

At times the leader sequence can contain elements for controlling gene expression by way of attenuation. The leader sequence will be translated and the availability of the encoded amino acids will affect the transcription of the rest of the mRNA.

It starts at the +1 position (where transcription begins) and ends one nucleotide before the start codon (usually AUG) of the coding region.

It usually contains a ribosome binding site (RBS), in bacteria also known as the Shine Dalgarno sequence (AGGAGGU). The 5' UTR may be a hundred or more nucleotides long, and the 3' UTR may be even longer (up to several kilobases in length).

An mRNA molecule codes for a protein through translation. The mRNA also contains regions that are not translated: in eukaryotes this includes the 5' untranslated region, 3' untranslated region, 5' cap and poly-A tail. In prokaryotic mRNA the 5' UTR is normally short. Some viruses and cellular genes have unusual long structured 5' UTRs which may have roles in gene expression.

Several regulatory sequences may be found in the 5' UTR:

- Binding sites for proteins, that may affect the mRNA's stability or translation, for example iron responsive elements, which occur in the 5' UTRs (and 3' UTRs) of a small number of eukaryotic mRNAs that regulate gene expression in response to iron.
- Regulatory elements that do not depend on proteins, such as riboswitches.
- Sequences that promote the initiation of translation.

### *3' UTR*

The three prime untranslated region (3' UTR) is a particular section of messenger RNA (mRNA). It follows the coding region.

An mRNA molecule codes for a protein through translation. The mRNA also contains regions that are not translated. In eukaryotes these regions are the cap, 5' untranslated region, 3' untranslated region, and polyA tail.

In prokaryotes mRNA structures have some differences as do histone mRNAs. However, both have 3' UTRs.

Several regulatory sequences are found in the 3' UTR:

- A polyadenylation signal, usually *AAUAAA*, or a slight variant. This marks the site of cleavage of the transcript approximately 30 base pairs past the signal, followed by the addition of several hundred adenine residues (poly-A tail).
- Binding sites for proteins, that may affect the mRNA's stability or location in the cell, like SECIS elements (which direct the ribosome to translate the codon UGA as selenocysteines rather than as a stop codon), or AU-rich elements (AREs), stretches consisting of mainly adenine and uracil nucleotides (which can either stabilize or destabilize the mRNA depending on the protein bound to it).
- Binding sites for miRNAs, a type of RNAi.

### Poly(A) Tail

The 3' poly(A) tail is a long sequence of adenine nucleotides (often several hundred) added to the 3' end of the pre-mRNA. This tail promotes export from the nucleus and translation, and protects the mRNA from degradation.

### Monocistronic versus Polycistronic mRNA

An mRNA molecule is said to be monocistronic when it contains the genetic information to translate only a single protein. This is the case for most of the eukaryotic mRNAs.

On the other hand, polycistronic mRNA carries the information of several genes, which are translated into several proteins. These proteins usually have a related function and are grouped and regulated together in an operon. Most of the mRNA found in bacteria and archea are polycistronic. Dicistronic or bicistronic is the term used to describe an mRNA that encodes only two proteins.

### mRNA Circularization

In eukaryotes it is thought that mRNA molecules form circular structures due to an interaction between the cap binding complex and poly(A)-binding protein. Circularization is thought to promote recycling of ribosomes on the same message leading to efficient translation.

### Degradation

Different mRNAs within the same cell have distinct lifetimes (stabilities). In bacterial cells, individual mRNAs can survive from seconds to more than an hour; in mammalian cells, mRNA lifetimes range from several minutes to days. The greater the stability of an mRNA, the more protein may be produced from that mRNA. The limited lifetime of mRNA enables a cell to alter protein synthesis rapidly in response to its changing needs. There are many mechanisms that lead to the destruction of a mRNA, some of which are described below.

### Prokaryotic mRNA Degradation

In prokaryotes the lifetime of mRNA is generally much shorter than in eukaryotes. Prokaryotes degrade messages by using a combination of ribonucleases, including endonucleases, 3' exonucleases, and 5' exonucleases. In some instances, small

RNA molecules (sRNA) tens to hundreds of nucleotides long can stimulate the degradation of specific mRNAs by base pairing with complementary sequences and facilitating ribonuclease cleavage. It was recently shown that bacteria also have a sort of 5' cap consisting of a triphosphate on the 5' end. Removal of two of the phosphates leaves a 5' monophosphate, causing the message to be destroyed by the endonuclease RNase E.

### *Eukaryotic mRNA Turnover*

Inside eukaryotic cells there is a balance between the processes of translation and mRNA decay. Messages that are being actively translated are bound by ribosomes, the eukaryotic initiation factors eIF-4E and eIF-4G, and poly(A)-binding protein. eIF-4E and eIF-4G block the decapping enzyme (DCP2), and poly(A)-binding protein blocks the exosome complex, protecting the ends of the message. The balance between translation and decay is reflected in the size and abundance of cytoplasmic structures known as P-bodies The poly(A) tail of the mRNA is shortened by specialized exonucleases that are targeted to specific messenger RNAs by a combination of cis-regulatory sequences on the RNA and trans-acting RNA-binding proteins. Poly(A) tail removal is thought to disrupt the circular structure of the message and destabilize the cap binding complex. The message is then subject to degradation by either the exosome complex or the decapping complex. In this way, translationally inactive messages can be destroyed quickly, while active messages remain intact.

### *AU-rich Element Decay*

The presence of AU-rich elements in some mammalian mRNAs tends to destabilize those transcripts through the action of cellular proteins that bind these sequences and stimulate poly(A) tail removal. Loss of the poly(A) tail is thought to promote mRNA degradation by facilitating attack by both the exosome complex and the decapping complex. Rapid mRNA degradation via AU-rich elements is a critical mechanism for preventing the overproduction of potent cytokines such as tumour necrosis factor (TNF) and granulocyte-macrophage colony stimulating factor (GM-CSF). AU-rich elements also regulate the biosynthesis of proto-oncogenic transcription factors like c-Jun and c-Fos.

### Nonsense Mediated Decay

Nonsense-mediated decay (NMD) is a cellular mechanism of mRNA surveillance that functions to detect nonsense mutations and prevent the expression of truncated or erroneous proteins. Following transcription, precursor mRNA undergoes an assemblage of ribonucleoprotein (RNP) components followed by regulatory pre-mRNA processing. Large average intron size in eukaryotic cells greatly increases the probability that aberrant mRNA splicing will result in the presence of a nonsense (stop) codon (UAA, UAG, UGA) somewhere within the open reading frame. NMD is triggered by exon junction complexes (EJCs) (components of the assembled RNP) that are deposited during pre-mRNA processing. While EJCs located in an open reading frame upstream of exon-exon junctions may facilitate ribosomal recruitment prior to displacement by a "pioneer" round of translation, EJCs located downstream of a nonsense codon are not displaced because the ribosome is released from the transcript before reaching it. These remaining EJCs function as tags for recruitment of UPF1 following the mRNA's transport out of the nucleus and into the cytosol where the RNA is degraded, for example by the exosome complex. NMD is not only a mechanism for degrading aberrant mRNA, however, as there are numerous examples of normal transcripts whose expression is regulated by this process including the plasticity protein Arc/Arg3.1.

Eukaryotic messages are subject to surveillance by nonsense mediated decay (NMD), which checks for the presence of premature stop codons (nonsense codons) in the message. These can arise via incomplete splicing, V(D)J recombination in the adaptive immune system, mutations in DNA, transcription errors, leaky scanning by the ribosome causing a frame shift, and other causes. Detection of a premature stop codon triggers mRNA degradation by 5' decapping, 3' poly(A) tail removal, or endonucleolytic cleavage.

# Chapter 9

# Techniques in Molecular Biology

Molecular biology is the branch of science which deals with the structure and function of macromolecules. It has various techniques which are also applied in other branches of biology.

Here are several techniques used in molecular biology which are also related to fields like genetics, recombinant DNA technology etc. These techniques include following:

***Expression cloning*** - One of the most widely used techniques of molecular biology to study protein function is expression cloning. In this technique DNA coding for a protein of interest is cloned. This is done by using PCR and/or restriction enzymes.

The DNA encoding the required protein is inserted into a plasmid which is known as an expression vector. This plasmid may include special promoter elements to drive production of the protein of interest, and may also have antibiotic resistance markers to help follow the plasmid.

***Polymerase Chain Reaction*** - is most frequently uses technique for copying DNA into millions of clones within few minutes. It can be used to insert restriction enzymes sites in a DNA fragment or to mutate a particular base sequence and also for determination about the presence of a particular DNA fragment in cDNA libraries.

***Gel electrophoresis*** - is used to separate fragments of DNA, RNA and proteins using different properties like according to size, ionic bond etc. For example agarose gel is used to separate RNA and DNA according to their size and SDS-PAGE is used to separate different proteins.

***Blotting techniques*** - is the technique of probing DNA(Southern blotting), RNA(Northern blotting), or proteins(Western blotting) molecules. Eastern blotting is used to recognize post translational modifications of proteins. Other well known and essential techniques in molecular biology include arrays like DNA microarray, allele specific oligonucleotide(ASO) etc.

## Role of Cellular Biology

Main tool used in cellular biology is microscope. A wide range of microscopes including Light microscope, Electron microscope, Fluorescent microscope are used for this purpose. The main purpose of cellular biology is to study different cellular organelles present in all types(prokaryotic-eukaryotic, plant-animal etc) of cells. These organelles include- nucleus, mitochondria, chloroplast, cytoplasm, endoplasmic reticulam, golgi apparatus, cell wall. cell membrane and many more cellular organelles.

Techniques involved in cellular biology other than microscopy include following:

1. Cell culture - in vitro technique of growing cells in laboratory.
2. Immunostaining - is the technique used to locate proteins present in the cell or parts of tissues.
3. Transfection - involve introduction of new gene into a cell
4. Cell fractionation - used to obtain cellular organelles for their study.

Other techniques used in cellular biology include centrifugation, flow cytometry, immunoprecipitation, gene knockdown etc.

## Polymerase Chain Reaction

Polymerase chain reaction is a technique of molecular biology which makes multiple copies of DNA from a single fragment. The process of polymerase chain reaction takes place as follows:

1. First the DNA molecule is denatured by heating under the temperature of 94 to 98 degree centigrade.

2. When the strands are separated, they are bound with DNA primers with the temperature 50 to 65 degree Celsius.
3. After annealing, the elongation of the separated DNA strands take place by using heat stable DNA polymerases like Taq (Thermus aquaticus).

There is another method of PCR named reverse transcription PCR which is used to amplify the RNA molecules, while real time PCR is used to measure the DNA o RNA molecule quantitatively.

### *Expression Cloning*

Cloning is the most useful technique of molecular biology because it is used to study the function of the proteins. This technique works in the way that it clones the DNA coding which makes protein, using the process of polymerase chain reaction or restriction enzymes and made multiple copies of that coding. The cloning converts the DNA samples into plasmid vectors. The plasmid either has promoter elements which make possible the production of protein of interest or they have the antibiotic resistant markers.

The plasmid produced from cloning can be inserted in the bacterial cell or the animal cell depending on the mode of experiment. If it is inserted in the bacterial cell, then three processes are involved which make possible the insertion of plasmid into the bacterial cell. These processes are transformation, conjugation and transduction. When the insertion of the cloned plasmid in the animal cell is concerned then the process of transfection is involved. Transfection techniques are of various types for example, microinjection, calcium phosphate transfection and liposome transfection.

But it is also possible that the plasmid can be inserted in the animal cell by bacteria or virus. If bacteria are involved for insertion then it is called as bactofention and mainly the specie of Agrobacterium tumafeciens is used for this purpose.

Plasmid can stay in the eukaryotic cell in two forms; either it mixes up with the genome of the host cell which results in the stable transfection or it stays independently in the cell by transient transfection. In both the cases, the protein of interest is in the cell and can be expressed. If the cell contains specific cell signaling factors or inducible promoters then protein is expressed efficiently than without these factors.

### *Gel Electrophoresis*

Gel electrophoresis is the basic technique of molecular biology which is used to separate DNA, RNA or protein molecules. The method involved is that, the molecules are placed in the immobilized gel and an electric current is passed through it. This current separates the molecules according to the size of the strands. Similarly proteins are also separated in the gel by using an SDS-PAGE gel or it can also be separated according to their size and the charge which they carry.

### *Allele Specific Oligonucleotide*

Allele specific technique is used to detect the single base mutations in the genes. It does not need any gel electrophoresis or polymerase chain reaction. In this technique, short probes are used usually 20 to 25 nucleotides long and they are also labeled. The labeled probes are exposed to the non-fragment target DNA; as a result hybridization takes place due to the short length probes. If there is only a single change in the nucleotide base, then the hybridization will stop. After this step, the target DNA is washed and those probes are removed which did not hybridize. Either radiation or florescence detects the presence of probe on the DNA fragment.

### *Arrays*

This is the technique of molecular biology which uses DNA arrays in the form of spots which are attached to the solid support. This solid support is usually microscopic slide which contains the single stranded DNA oligonucleotide fragment present on each spot.

This technique is very feasible because it contains large amount of very short spots with DNA fragments on each slide. The spot is so small that it contains a single DNA fragment. This fragment is complementary to a single DNA sequence. The process starts with the extraction of RNA molecule from a tissue which is converted in the form of cDNA. The DNA fragments present on the spots in the slide are then hybridized with these cDNAs. They can be visualized after hybridization.

This technique is useful in the gene expression data as many arrays can be made with the same position of the DNA fragments on the spots, so that the gene expression of healthy and diseases genes can be compared. For example common yeast consists of 7000 genes. With the microarray, it is possible to observe the expression of each gene qualitatively. It is also possible to view the changes in the expression of these genes due to temperature. There are also other methods of construction of arrays instead of suing DNA fragments. For example an antibody array can be used to determine the presence f proteins or bacteria in the sample of blood.

## Protein Folding and Modifications

A polypeptide chain is synthesized on large cellular structures, the ribosomes, by a complex process in which assembly of amino acids in a particular sequence is dictated by messenger RNA (mRNA). The nascent polypeptide chain undergoes folding and, in many cases, chemical modification to generate the final protein.

Any polypeptide chain containing $n$ residues could, in principle, fold into $8^n$ conformations. This value is based on the fact that only eight bond angles are stereochemically allowed in the polypeptide backbone. In general, however, all molecules of any protein species adopt a single conformation, called the *native state,* which is the most stably folded form of the molecule. Misfolding to non-native conformations is suppressed by two distinct mechanisms. At the molecular level, a protein folds through a pathway that favors only a few intermediate steps. Furthermore, a cellular system prevents misfolded proteins from forming. After a protein has carried out its functions, specific sequences that limit the life span of the protein target it for degradation.

## The Information for Protein Folding is Encoded in the Sequence

The realization that the amino acid sequence of a protein determines its folding came from in vitro studies on protein unfolding and refolding. Thermal energy from heat, extremes of pH that alter the charges on amino acid side chains, and chemicals such as urea or guanidine hydrochloride at concentrations of 6–8 M can disrupt the weak noncovalent bonds that stabilize the native conformation of a protein. The denaturation resulting from such treatment causes a protein to lose both its compact conformation and activity. Most denatured proteins precipitate in solution because hydrophobic groups, normally buried inside the molecules, interact with similar regions of other unfolded molecules, causing them to form an insoluble aggregate.

Many proteins that are completely unfolded in 8 M urea and β-mercaptoethanol (which reduces disulfide bonds) can *renature* (refold) into their native state when the denaturing reagents are removed by dialysis. During renaturation, all the disulfide, hydrogen, and hydrophobic bonds that stabilize the native conformation are re-formed. Thus, in this case proteins can be carried through a denaturationrenaturation cycle, which first destroys and then reestablishes their original structure and function. Because renaturation requires no cofactors or other proteins, at least in the test tube, protein folding is a self-assembly process.

Treatment with an 8 M urea solution containing mercaptoethanol ($HSCH_2CH_2OH$) completely denatures most proteins. The urea breaks intramolecular hydrogen and hydrophobic bonds, and the mercaptoethanol reduces each disulfide bridge (–S–S–) to two sulfhydryl (–SH) groups. When these chemicals are removed by dialysis, the –SH groups on the unfolded chain oxidize spontaneously to reform disulfide bridges, and the polypeptide chain simultaneously refolds into its native conformation. The observation by Christian Anfinsen of such reversible denaturation and renaturation of ribonuclease, an enzyme that degrades RNA, provided a clue that the information for folding a protein lies in its sequence. A general mechanism by which proteins refold in vitro has been elucidated in experiments in which the renaturing conditions are carefully adjusted and the refolding reaction is interrupted at various time intervals. Such studies have shown that the polypeptide goes through several transient reconfigurations, including a "molten globule" state, before the native tertiary conformation is reached. In the case of ribonuclease, which has several internal disulfide bonds, the folding pathway involves rearrangements of disulfide bond pairs to the native conformation.

Under appropriate refolding conditions, the molecule condenses around a hydrophobic core into a compact, but non-native, intermediate, called a *molten globule.*In this folding intermediate, much of the secondary structure is present. Long-range interactions then form the tertiary structure, folding the molecule into its native three-dimensional conformation.

## Folding of Proteins in Vivo is Promoted by Chaperones

Folding of proteins in vitro is an inefficient process, with only a minority of unfolded molecules undergoing complete folding within a few minutes. Clearly, in vivo most protein molecules must rapidly fold into their correct shape; otherwise, cells would waste much energy in the synthesis of nonfunctional proteins and in the degradation of misfolded or unfolded proteins. More than 95 percent of the proteins present within cells have been shown to be in their native conformation, despite high protein concentrations (H–100 mg/ml), which usually cause proteins to precipitate in vitro.

The explanation for the cell's remarkable efficiency in promoting protein folding probably lies in chaperones, a family of proteins found in all organisms from bacteria to humans. Chaperones are located in every cellular compartment, bind a wide range of proteins, and may be part of a general protein-folding mechanism. There are two general families of chaperones: *molecular chaperones,* which bind and stabilize unfolded or partially folded proteins, thereby preventing these proteins from being degraded; and *chaperonins,* which directly facilitate their folding.

Chaperones have ATPase activity, and their ability to bind and stabilize their target proteins is specific and dependent on ATP hydrolysis. Binding of chaperones to partially folded proteins suggests that the folding process could be regulated at intermediate steps. Molecular chaperones consist of the Hsp70 family of proteins, which includes Hsp70 in the cytosol and mitochondrial matrix, Bip in the endoplasmic reticulum, and DnaK, a bacterial chaperone. First identified by its rapid appearance after a cell has been stressed by heat shock, Hsp70 is the major chaperone protein in all organisms. When bound to ATP, Hsp70 assumes an open form in which an exposed hydrophobic pocket transiently binds to exposed hydrophobic regions of the unfolded target protein. Hydrolysis of the bound ATP causes Hsp70 to assume a closed form, releasing the target protein. Molecular chaperones are thought to bind all nascent polypeptide chains as they are being synthesized on ribosomes. In bacteria, 85 percent of the proteins are released from their chaperone and go on to fold normally; an even higher percentage of proteins in eukaryotes follow this pathway.

(a) Many proteins 1 fold into their proper three-dimensional structure with the assistance of Hsp70, a molecular chaperone that transiently binds to a nascent polypeptide as it emerges from a ribosome. Proper folding of some proteins 2 also depends on the chaperonin TCiP, a large barrel-shaped complex of Hsp60 units.

(b) GroEL, the bacterial homolog of TCiP, is a barrel-shaped complex of 14 identical 60,000-MW subunits arranged in two stacked rings. In the absence of ATP or presence of ADP, GroEL exists in a "tight" conformational state that binds partially folded or misfolded proteins. Binding of ATP shifts GroEL to a more open, "relaxed" state, which releases the folded protein.

Proper folding of a small proportion of proteins (e.g., the cytoskeletal proteins actin and tubulin) requires additional assistance, which is provided by chaperonins. Eukaryotic chaperonins, called *TCiP,* are large, barrel-shaped, multimeric complexes composed of eight Hsp60 units.

The bacterial homolog, known as *GroEL,* contains 14 identical subunits. The GroEL folding mechanism, which is better understood than TCiP-mediated folding, serves as a reasonable general model. In bacteria, a partially folded or misfolded polypeptide is inserted into the cavity of GroEL, where it binds to the inner wall and folds into its native conformation. In an ATP-dependent step, GroEL expands and the protein exits GroEL, a process assisted by a co-chaperonin, GroES, which caps the ends of GroEL. Because the eukaryotic chaperonin TCiP lacks a GroES-type co-chaperonin, the last step must differ in eukaryotes. Moreover, the size of the cavity in TCiP limits this folding pathway to polypeptides smaller than 55 kDa.

## Chemical Modifications and Processing Alter the Biological Activity of Proteins

Nearly every protein in a cell is chemically altered after its synthesis on a ribosome. Such modifications may alter the activity, life span, or cellular location of proteins, depending on the nature of the alteration. Protein alterations fall into two categories: chemical modification and processing. *Chemical modification* involves the linkage of a chemical group to the terminal amino or carboxyl groups or to reactive groups in the side chains of internal residues; in some cases, these modifications are reversible. *Processing* involves the removal of peptide segments and generally is irreversible. *Acetylation,* the addition of an acetyl group ($CH_3CO$) to the amino group of the N-terminal residue is the most common form of chemical modification, involving an estimated 80 percent of all proteins:

This modification may play an important role in controlling the life span of proteins within cells, as nonacetylated proteins are rapidly degraded by intracellular proteases. As discussed later, residues at or near the termini of some membrane proteins are chemically modified by addition of long lipidlike groups. Attachment of these hydrophobic "tails," which function to anchor proteins to the lipid bilayer, constitute one way that cells restrict certain proteins to membranes. The internal residues in proteins can be modified by attachment of a variety of chemical groups to their side chains. The most important modification is *phosphorylation* of serine, threonine, and tyrosine residues. We will encounter numerous examples of proteins whose activity is regulated by reversible phosphorylation and dephosphorylation.

The side chains of asparagine, serine, and threonine are sites for *glycosylation,* the attachment of linear and branched carbohydrate chains. Many secreted proteins and membrane proteins contain glycosylated residues. Various less common modifications are found in a limited number of proteins. Examples of modified internal residues produced by hydroxylation, methylation, and carboxylation. These modifications occur after synthesis of the polypeptide chain.

These modifications occur after synthesis of the polypeptide chain. Unlike chemical modification of residues, which often is reversible, processing of some proteins causes irreversible changes that alter their activity. In the most common form of processing, residues are removed from the C- or N-terminus of a polypeptide by cleavage of the peptide bond in a reaction catalyzed by proteases. Proteolytic cleavage is a common mechanism of activation or inactivation, especially of enzymes involved in blood coagulation or digestion. As discussed later, the activity of certain digestive enzymes is controlled by this mechanism. Proteolysis also generates active peptide hormones, such as EGF mentioned earlier and insulin, from larger precursor polypeptides.

An unusual type of processing, termed *protein selfsplicing,* occurs in bacteria and primitive eukaryotes. *Splicing* refers to a process analogous to editing film: an internal segment of polypeptide, an intein, is removed and the ends of the polypeptide are rejoined. Unlike proteolytic processing, protein self-splicing is an autocatalytic process, which proceeds by itself without the involvement of enzymes. The excised peptide appears to eliminate itself from the protein by a mechanism similar to that used in processing of RNA molecules. In vertebrate cells, processing of some proteins involves self-cleavage, but the subsequent ligation step is absent. One such protein is Hedgehog, which is critical to a number of developmental processes.

A segment (red) of a polypeptide, called an *intein,* is removed by cleavage of two peptide bonds (red arrows), leaving two segments (blue and green). One new peptide bond then forms between the two segments (blue arrow), regenerating a continuous polypeptide backbone. The process is autocatalytic and does not depend on enzymes. Consensus splice sites in the polypeptide chain mark the points of breakage and re-formation of the backbone. The excised segment is thought to be exposed at the surface of the folded protein.

## Cells Degrade Proteins Via Several Pathways

Cells have both extracellular and intracellular pathways for degrading proteins. The major extracellular pathway is the system of *digestive proteases,* which break down ingested proteins to polypeptides in the intestinal tract. These include *endoproteases* such as trypsin and chymotrypsin, which cleave the protein backbone adjacent to basic and aromatic residues;

*exopeptidases,* which sequentially remove residues from the N-terminus (aminopeptidases) or C-terminus (carboxypeptidases) of proteins; and *peptidases,* which split oligopeptides into di- and tripeptides and individual amino acids. These small molecules then are transported across the intestinal lining into the bloodstream.

The life span of intracellular proteins varies from as short as a few minutes for mitotic cyclins, which help regulate passage through mitosis, to as long as the age of an organism for proteins in the lens of the eye. Cells have several intracellular proteolytic pathways for degrading misfolded or denatured proteins, normal proteins whose concentration must be decreased, and foreign proteins taken up by the cell. One major intracellular pathway involves degradation by enzymes within lysosomes, membrane-limited organelles whose interior is acidic. Distinct from the lysosomal pathway are cytosolic mechanisms for degrading proteins. The best-understood pathway, the ubiquitin-mediated pathway, involves two steps: addition of a chain of ubiquitin molecules to an internal lysine side chain of a target protein and proteolysis of the ubiquitinated protein by a proteasome, a large, cylindrical multisubunit complex. The numerous proteasomes present in the cell cytosol proteolytically cleave ubiquitin-tagged proteins in an ATP-dependent process that yields peptides and intact ubiquitin molecules.

A conjugating enzyme catalyzes formation of a peptide bond between ubiquitin (Ub) and the side-chain $-NH_2$ of a lysine residue in a target protein. Additional Ub molecules are added, forming a multiubiquitin chain. This chain is thought to direct the tagged protein to a proteasome, which cleaves the protein into numerous small peptide fragments. (b) Computer-generated image reveals cylindrical structure of a proteasome with a cap at each end of a core. Proteolysis of ubiquitin-tagged proteins occurs along the inner wall of the core. To be targeted for degradation by the ubiquitinmediated pathway, a protein must contain a structure that is recognized by a ubiquitinating enzyme complex. Different conjugating enzymes recognize different degradation signals in target proteins. For example, the internal sequence Arg-X-X-Leu-Gly-X-Ile-Gly-Asx in mitotic cyclin is recognized by the ubiquitin-conjugating enzyme E1. Internal sequences enriched in proline, glutamic acid, serine, and threonine (PEST sequences) are recognized by other enzymes. The life span of many cytosolic proteins is correlated with the identity of the N-terminal residue, suggesting that certain residues at the N-terminus favor rapid ubiquitination. For example, short-lived proteins that are degraded within 3 minutes in vivo commonly have Arg, Lys, Phe, Leu, or Trp at their N-terminus. In contrast, a stabilizing amino acid such as Cys, Ala, Ser, Thr, Gly, Val, or Met is present at the N-terminus in long-lived proteins that resist proteolytic attack for more than 30 hours. All newly synthesized proteins have methionine, a stabilizing amino acid, at the N-terminus. Thus subsequent enzymatic alteration that generates one of the destabilizing amino acids at the N-terminus is necessary to target a protein for degradation.

## Aberrantly Folded Proteins are Implicated in Slowly Developing Diseases

As noted earlier, each protein species normally folds into a single, energetically favorable conformation that is specified by its amino acid sequence. Recent evidence suggests, however, that a protein may fold into an alternative three- dimensional structure for reasons that have not yet been identified. Such "misfolding" not only leads to a loss of the normal function of a protein but also marks it for proteolytic degradation. The subsequent accumulation of proteolytic fragments contributes to certain degenerative diseases characterized by the presence of insoluble protein plaques in various organs including the liver and brain.

Alzheimer's disease, for example, is marked by formation of plaques and tangles in a deteriorating brain. The filaments composing these structures are derived from proteolytic products of abundant natural proteins such as amyloid precursor protein, a transmembrane protein, and Tau, a microtubule-binding protein. Plaques in other organs are formed from proteolytic fragments of natural proteins such as gelsolin, an actin-binding protein, and serum albumin, a blood protein. The polypeptide fragments liberated by proteolysis polymerize into very stable filaments. A degeneration of the brain, similar to that seen in Alzheimer's disease, is thought to be caused by *prions,* an infectious protein agent derived by proteolysis and re-folding of a normal brain protein. The amyloid plaque in the brain of an Alzheimer's patient appears as a tangle of filaments. (b) In the atomic force microscope, the filaments are seen to be regular arrangements of a short 47-residue fragment, called β-*amyloid peptide,* produced by proteolysis of amyloid precursor protein.

- The amino acid sequence of a protein dictates its folding into a specific three-dimensional conformation, the native state.
- Folding of denatured proteins in vitro proceeds through intermediates having secondary and non-native tertiary structure.
- Protein folding in vivo occurs with the assistance of two types of special proteins. Molecular chaperones (Hsp70 proteins) bind to nascent polypeptides emerging from ribosomes and prevent their misfolding. Chaperonins, large complexes of Hsp60-like proteins, shelter some partially folded or misfolded proteins in a barrel-like cavity, providing additional timc for proper folding.
- Following their synthesis, all proteins are modified in various ways that alter their structure and function.
- The life span of intracellular proteins is largely determined by their susceptibility to proteolytic degradation by various pathways.
- The presence of certain internal sequences or Nterminal residues targets cytosolic proteins for addition of ubiquitin and subsequent proteolysis within a proteasome.

## Genomics and Proteomics

Genomics is a discipline in genetics concerning the study of the genomes of organisms. The field includes intensive efforts to determine the entire DNA sequence of organisms and fine-scale genetic mapping efforts. The field also includes studies of intragenomic phenomena such as heterosis, epistasis, pleiotropy and other interactions between loci and alleles within the genome. In contrast, the investigation of the roles and functions of single genes is a primary focus of molecular biology or genetics and is a common topic of modern medical and biological research. Research of single genes does not fall into the definition of genomics unless the aim of this genetic, pathway, and functional information analysis is to elucidate its effect on, place in, and response to the entire genome's networks.

For the United States Environmental Protection Agency, "the term "genomics" encompasses a broader scope of scientific inquiry associated technologies than when genomics was initially considered. A genome is the sum total of all an individual organism's genes. Thus, genomics is the study of all the genes of a cell, or tissue, at the DNA (genotype), mRNA (transcriptome), or protein (proteome) levels."

### *History*

The first genomes to be sequenced were those of a virus and a mitochondrion, and were done by Fred Sanger. His group established techniques of sequencing, genome mapping, data storage, and bioinformatic analyses in the 1970-1980s. A major branch of genomics is still concerned with sequencing the genomes of various organisms, but the knowledge of full genomes has created the possibility for the field of functional genomics, mainly concerned with patterns of gene expression during various conditions.

The most important tools here are microarrays and bioinformatics. Study of the full set of proteins in a cell type or tissue, and the changes during various conditions, is called proteomics. A related concept is materiomics, which is defined as the study of the material properties of biological materials (e.g. hierarchical protein structures and materials, mineralized biological tissues, etc.) and their effect on the macroscopic function and failure in their biological context, linking processes, structure and properties at multiple scales through a materials science approach. The actual term 'genomics' is thought to have been coined by Dr. Tom Roderick, a geneticist at the Jackson Laboratory (Bar Harbor, ME) over beer at a meeting held in Maryland on the mapping of the human genome in 1986.

In 1972, Walter Fiers and his team at the Laboratory of Molecular Biology of the University of Ghent (Ghent, Belgium) were the first to determine the sequence of a gene: the gene for Bacteriophage MS2 coat protein. In 1976, the team determined the complete nucleotide-sequence of bacteriophage

MS2-RNA. The first DNA-based genome to be sequenced in its entirety was that of bacteriophage Φ-X174; (5,368 bp), sequenced by Frederick Sanger in 1977. The first free-living organism to be sequenced was that of *Haemophilus influenzae* (1.8 Mb) in 1995, and since then genomes are being sequenced at a rapid pace. As of September 2007, the complete sequence was known of about 1879 viruses, 577 bacterial species and roughly 23 eukaryote organisms, of which about half are fungi. Most of the bacteria whose genomes have been completely sequenced are problematic disease-causing agents, such as *Haemophilus influenzae*. Of the other sequenced species, most were chosen because they were well-studied model organisms or promised to become good models. Yeast (*Saccharomyces cerevisiae*) has long been an important model organism for the eukaryotic cell, while the fruit fly *Drosophila melanogaster* has been a very important tool (notably in early pre-molecular genetics). The worm *Caenorhabditis elegans* is an often used simple model for multicellular organisms. The zebrafish *Brachydanio rerio* is used for many developmental studies on the molecular level and the flower *Arabidopsis thaliana* is a model organism for flowering plants. The Japanese pufferfish (*Takifugu rubripes*) and the spotted green pufferfish (*Tetraodon nigroviridis*) are interesting because of their small and compact genomes, containing very little non-coding DNA compared to most species. The mammals dog (*Canis familiaris*), brown rat (*Rattus norvegicus*), mouse (*Mus musculus*), and chimpanzee (*Pan troglodytes*) are all important model animals in medical research.

### Human Genomics

A rough draft of the human genome was completed by the Human Genome Project in early 2001, creating much fanfare. By 2007 the human sequence was declared "finished" (less than one error in 20,000 bases and all chromosomes assembled). Display of the results of the project required significant bioinformatics resources. The sequence of the human reference assembly can be explored using the UCSC Genome Browser.

### Bacteriophage Genomics

Bacteriophages have played and continue to play a key role in bacterial genetics and molecular biology. Historically, they were used to define gene structure and gene regulation.

Also the first genome to be sequenced was a bacteriophage. However, bacteriophage research did not lead the genomics revolution, which is clearly dominated by bacterial genomics. Only very recently has the study of bacteriophage genomes become prominent, thereby enabling researchers to understand the mechanisms underlying phage evolution. Bacteriophage genome sequences can be obtained through direct sequencing of isolated bacteriophages, but can also be derived as part of microbial genomes. Analysis of bacterial genomes has shown that a substantial amount of microbial DNA consists of prophage sequences and prophage-like elements. A detailed database mining of these sequences offers insights into the role of prophages in shaping the bacterial genome.

### *Cyanobacteria Genomics*

At present there are 24 cyanobacteria for which a total genome sequence is available. 15 of these cyanobacteria come from the marine environment. These are six *Prochlorococcus* strains, seven marine *Synechococcus* strains, *Trichodesmium erythraeum* IMS101 and *Crocosphaera watsonii* WH8501. Several studies have demonstrated how these sequences could be used very successfully to infer important ecological and physiological characteristics of marine cyanobacteria.

However, there are many more genome projects currently in progress, amongst those there are further *Prochlorococcus* and marine *Synechococcus* isolates, *Acaryochloris* and *Prochloron*, the $N_2$-fixing filamentous cyanobacteria *Nodularia spumigena*, *Lyngbya aestuarii* and *Lyngbya majuscula*, as well as bacteriophages infecting marine cyanobaceria.

Thus, the growing body of genome information can also be tapped in a more general way to address global problems by applying a comparative approach. Some new and exciting examples of progress in this field are the identification of genes for regulatory RNAs, insights into the evolutionary origin of photosynthesis, or estimation of the contribution of horizontal gene transfer to the genomes that have been analysed.

## Proteomics

Proteomics is the large-scale study of proteins, particularly their structures and functions. Proteins are vital parts of living organisms, as they are the main components of the physiological metabolic pathways of cells. The term "proteomics" was first coined in 1997 to make an analogy with genomics, the study of the genes. The word "proteome" is a blend of "protein" and "genome", and was coined by Marc Wilkins in 1994 while working on the concept as a PhD student. The proteome is the entire complement of proteins, including the modifications made to a particular set of proteins, produced by an organism or system. This will vary with time and distinct requirements, or stresses, that a cell or organism undergoes.

### Complexity of the Problem

After genomics, proteomics is considered the next step in the study of biological systems. It is much more complicated than genomics mostly because while an organism's genome is more or less constant, the proteome differs from cell to cell and from time to time. This is because distinct genes are expressed in distinct cell types. This means that even the basic set of proteins which are produced in a cell needs to be determined.

In the past this was done by mRNA analysis, but this was found not to correlate with protein content. It is now known that mRNA is not always translated into protein, and the amount of protein produced for a given amount of mRNA depends on the gene it is transcribed from and on the current physiological state of the cell. Proteomics confirms the presence of the protein and provides a direct measure of the quantity present.

### Post-translational Modifications

Not only does the translation from mRNA cause differences, many proteins are also subjected to a wide variety of chemical modifications after translation. Many of these post-translational modifications are critical to the protein's function.

### Phosphorylation

One such modification is phosphorylation, which happens to many enzymes and structural proteins in the process of cell signaling. The addition of a phosphate to particular amino acids—most commonly serine and threonine mediated by serine/threonine kinases, or more rarely tyrosine mediated by tyrosine kinases—causes a protein to become a target for binding or interacting with a distinct set of other proteins that recognize the phosphorylated domain.

Because protein phosphorylation is one of the most-studied protein modifications, many "proteomic" efforts are geared to determining the set of phosphorylated proteins in a particular cell or tissue-type under particular circumstances. This alerts the scientist to the signaling pathways that may be active in that instance.

### *Ubiquitination*

Ubiquitin is a small protein that can be affixed to certain protein substrates by enzymes called E3 ubiquitin ligases. Determining which proteins are poly-ubiquitinated can be helpful in understanding how protein pathways are regulated. This is therefore an additional legitimate "proteomic" study. Similarly, once it is determined what substrates are ubiquitinated by each ligase, determining the set of ligases expressed in a particular cell type will be helpful.

### *Additional Modifications*

Listing all the protein modifications that might be studied in a "Proteomics" project would require a discussion of most of biochemistry; therefore, a short list will serve here to illustrate the complexity of the problem. In addition to phosphorylation and ubiquitination, proteins can be subjected to (among others) methylation, acetylation, glycosylation, oxidation and nitrosylation. Some proteins undergo ALL of these modifications, often in time-dependent combinations, aptly illustrating the potential complexity one has to deal with when studying protein structure and function.

### *Distinct Proteins are Made under Distinct Settings*

Even if one is studying a particular cell type, that cell may make different sets of proteins at different times, or under different conditions. Furthermore, as mentioned, any one protein can undergo a wide range of post-translational modifications. Therefore a "proteomics" study can become quite complex very quickly, even if the object of the study is very restricted. In more ambitious settings, such as when a biomarker for a tumour is sought - when the proteomics scientist is obliged to study sera samples from multiple cancer patients - the amount of complexity that must be dealt with is as great as in any modern biological project.

## Limitations to Genomic Study

Scientists are very interested in proteomics because it gives a much better understanding of an organism than genomics. First, the level of transcription of a gene gives only a rough estimate of its level of expression into a protein. An mRNA produced in abundance may be degraded rapidly or translated inefficiently, resulting in a small amount of protein. Second, as mentioned above many proteins experience post-translational modifications that profoundly affect their activities; for example some proteins are not active until they become phosphorylated.

Methods such as phosphoproteomics and glycoproteomics are used to study post-translational modifications. Third, many transcripts give rise to more than one protein, through alternative splicing or alternative post-translational modifications. Fourth, many proteins form complexes with other proteins or RNA molecules, and only function in the presence of these other molecules. Finally, protein degradation rate plays an important role in protein content.

## Methods of Studying Proteins

### *Determining Proteins which are Post-translationally Modified*

One way in which a particular protein can be studied is to develop an antibody which is specific to that modification. For example, there are antibodies which only recognize certain proteins when they are tyrosine-phosphorylated, known as phospho-specific antibodies; also, there are antibodies specific to other modifications. These can be used to determine the set of proteins that have undergone the modification of interest.

For sugar modifications, such as glycosylation of proteins, certain lectins have been discovered which bind sugars. These too can be used.

A more common way to determine post-translational modification of interest is to subject a complex mixture of proteins to electrophoresis in "two-dimensions", which simply means that the proteins are electrophoresed first in one direction, and then in another... this allows small differences in a protein to be visualized by separating a modified protein from its unmodified form. This methodology is known as "two-dimensional gel electrophoresis". Recently, another approach has been developed called PROTOMAP which combines SDS-PAGE with shotgun proteomics to enable detection of changes in gel-migration such as those caused by proteolysis or post translational modification.

### Determining the Existence of Proteins in Complex Mixtures

Classically, antibodies to particular proteins or to their modified forms have been used in biochemistry and cell biology studies. These are among the most common tools used by practicing biologists today. For more quantitative determinations of protein amounts, techniques such as ELISAs can be used. For proteomic study, more recent techniques such as matrix-assisted laser desorption/ionization (MALDI) have been employed for rapid determination of proteins in particular mixtures and increasingly electrospray ionization (ESI).

### Computational Methods in Studying Protein Biomarkers

Computational predictive models have shown that extensive and diverse feto-maternal protein trafficking occurs during pregnancy and can be readily detected non-invasively in maternal whole blood. This computational approach circumvented a major limitation, the abundance of maternal proteins interfering with the detection of fetal proteins, to fetal proteomic analysis of maternal blood. Computational models can use fetal gene transcripts previously identified in maternal whole blood to create a comprehensive proteomic network of the term neonate. Such work shows that the fetal proteins detected in pregnant woman's blood originate from a diverse group of tissues and organs from the developing fetus. The proteomic networks contain many biomarkers that are proxies for development and illustrate the potential clinical application of this technology as a way to monitor normal and abnormal fetal development. An information theoretic framework has also been introduced for biomarker discovery, integrating biofluid and tissue information. This new approach takes advantage of functional synergy between certain biofluids and tissues with the potential for clinically significant findings not possible if tissues and biofluids were considered individually. By conceptualizing tissue-biofluid as information channels, significant biofluid proxies can be identified and then used for guided development of clinical diagnostics. Candidate biomarkers are then predicted based on information transfer criteria across the tissue-biofluid channels. Significant biofluid-tissue relationships can be used to prioritize clinical validation of biomarkers.

## Establishing Protein-protein Interactions

Most proteins function in collaboration with other proteins, and one goal of proteomics is to identify which proteins interact. This is especially useful in determining potential partners in cell signaling cascades. Several methods are available to probe protein-protein interactions. The traditional method is yeast two-hybrid analysis. New methods include protein microarrays, immunoaffinity chromatography followed by mass spectrometry, dual polarisation interferometry, Microscale Thermophoresis and experimental methods such as phage display and computational methods

### *Practical Applications of Proteomics*

One of the most promising developments to come from the study of human genes and proteins has been the identification of potential new drugs for the treatment of disease. This relies on genome and proteome information to identify proteins associated with a disease, which computer software can then use as targets for new drugs. For example, if a certain protein is implicated in a disease, its 3D structure provides the information to design drugs to interfere with the action of the protein. A molecule that fits the active site of an enzyme, but cannot be released by the enzyme, will inactivate the enzyme. This is the basis of new drug-discovery tools, which aim to find new drugs to inactivate proteins involved in disease. As genetic differences among individuals are found, researchers expect to use these techniques to develop personalized drugs that are more effective for the individual.

# Bibliography


Ausubel, F.M. : *Current Protocols in Molecular Biology,* New York: John Wiley and Sons, 2009.

Brodwin, Paul : *Biotechnology and Culture: Bodies, Anxieties, Ethics,* Bloomington: Indiana University Press, 2023.

Currah L. and Proctor F. J.: *Onions in Tropical Regions*, Kent, Natural Resources Institute, 2010.

Dudley, E.: *The Critical Villager: Beyond Community Participation*, London, Routledge, 2013.

E. Ramann: *The Evolution and Classification of Soils*, Asiatic Pub, Calcutta, 2016.

Elliott, B N : *Biochemistry and Molecular Biology*, Oxford University Press, Delhi, 2015.

Fumento, Michael: *Bioevolution: How Biotechnology is Changing Our World*, San Francisco, Encounter Books, 2013.

Graf, Alfred Byrd: *Advances in Plant Physiology*, Rajat Pub, Delhi, 2018.

Hardy B.: *Biology and Agronomy of Forage Arachis*, Cali, International Centre for Tropical Agriculture, 2014.

Jeffers P.: *Evaluation of Four Onion Varieties in Montserrat*, Plymouth, CARDI, 2012.

Jones, R. M.: *Plant Resources of South-East Asia,* Wageningen, Pudoc Scientific Publishers, 2012.

Khanna, V K : *Objective Genetics, Biotechnology, Biochemistry and Forestry,* I.K. International Publishing House, Delhi, 2018.

Kuppuram, G & K. Kumudamani: *History of Science and Technology in India,* Delhi, Sundeep Prakashan, 2010.

Kurzweil, Ray: *The Age of Spiritual Machines*, New York, Penguin Books, 2019.

Larry V. McIntire : *Biotechnology: Science, Engineering, and Ethical Challenges for the Twenty-first Century,* Washington, DC: Joseph Henry Press, 2016.

Meena Francis: *Biotech's Dictionary of Biochemistry*, Biotech Books, Delhi, 2020.

Muneesh Kainth: *Chordate Embryology*, Dominant, Delhi, 2023.

Nobel, P. S.: *Physicochemical and Environmental Plant Physiology*, Academic Press, San Diego, 2019.

Old, R.W. : *Principles of Gene Manipulation*, London, Blackwell Scientific Publications, 2009.

Oldham P.: *Cost of Production of Major Tree Crops in Dominica*, Roseau, Ministry of Agriculture, 2011.

Pemberton, R. W.: *Predictable Risk to Native Plants in Weed Biological Control,* Oecologia, 2017.

Qystein V. Sjaastad: *Physiology of Domestic Animals*, International Book Distributing Co., Delhi, 2015.

Ragone D.: *Breadfruit: Artocarpus Altilis (Parkinson) Fosberg*, Rome, International Plant Genetic Resources Institute, 2007.

Rifkin, Jeremy: *The Biotech Century*, New York, Penguin Putnam, 2018.

Rutherford Lyn.: *A Gourmet's Book of Mushrooms & Truffles*, Sydney, Golden Press Pvt. Ltd., 2011.

Sharma, Pradeep : *Biochemistry and Organisation of Cells*, RBSA Pub, Delhi, 2016.

Stover, R. H. and Simmonds N. W.: *Bananas*, United Kingdom: Longman Scientific and Technical, 2011.

Swarnim, K. : *A Textbook of Biochemistry and Microbiology*, Surendra Pub, Delhi, 2020.

Tawde, A. B.: *Propagation and Rootstocks of Mango*, New Delhi, Malhotra, 2013.

Urton, Gary: *The Social Life of Numbers*, Austin, University of Texas Press, 2017.

Vanangamudi, K.: *Principles and Methods of Plant Breeding*, International Book, Delhi, 2015.

Whealy K.: *The Garden Seed Inventory*, Decorah, Seed Saver Publications, 2008.

White, G.F.: *Natural Hazards: Local, National, Global*, Oxford University Press, New York, 1994.

Woolfe Jennifer A.: *The Potato in the Human Diet*, Cambridge, Cambridge University Press, 2009.

Yadav, M.: *Nutritional Biochemistry and Metabolism*, Arise Pub, Delhi, 2018.

# Index


A

Acids, 147, 150, 163, 204, 205.
Acrosome Reaction, 23.
Action Spectra, 64.
Amino Acid Synthesis, 168.
Amino Acids, 132, 169, 171.
AmpFLP, 155.
Anatomy, 22.
Artificial Ribozymes, 253.
ATP Synthase, 44, 70.
Autosomal Recessive, 177.

B

Bacteriophage Genomics, 286.
Bioinformatics, 164.
Biomarkers, 291.
Biomembranes, 105.
Biomolecules, 127, 128, 171.
Blood Group Antigens, 106.
Breast Cancer, 184.

C

Cell Cycle, 1, 8, 11.
Cell Division, 1, 5.
Cell Growth, 1, 2.
Cell growth Disorders, 7.
Cell Populations, 2.
Cell populations, 2.
Cellular Organization, 103.
Checkpoints, 14, 206.
Chemical Damage, 220.
Chlorophyll Biosynthesis, 74.
Chloroplasts, 65, 85, 88, 110.
Cholesterol and Steroids, 106.
Chromosome Banding, 107.
Colon Cancer, 181, 182, 183.
Corona Radiata, 22.
Cortical Reaction, 23.
Cyanobacteria Genomics, 287.
Cytoplasmic Organelles, 109.

D

Deadenylation, 257.
Direct Reversal, 202.
DNA, 1, 2, 4, 5, 286, 287.
DNA Damage, 196, 197, 200, 201, 206, 208.
DNA Databases, 158.
DNA Profiling, 151, 152.
DNA Replication, 195.
DNA-binding Proteins, 145.
DNA-modifying Enzymes, 147.
Down Syndrome, 187, 188, 189, 246.
Downstream Effects, 256.
Drugs, 128.

E

Elongation, 171, 172, 173, 227.
Eukaryotic Cells, 105.
Expression Cloning, 274.

F

Fertilization, 17, 22.
Forensics, 151.
Fusion, 20, 21, 23.

G

Gel Electrophoresis, 275.
Genes and Genomes, 143.

Genomics and Proteomics, 284.
Glyoxisomes, 111.
Golgi Apparatus, 111.

**H**

Human Genomics, 286.
Hypothyroidism, 180, 181.

**I**

Ionic Bonding, 115.
Isoenzymes, 133.

**K**

Karyotype, 107, 108.

**L**

Lignin, 131.
Lysosomes, 111.

**M**

Mammals, 3, 4, 21.
Mechanism, 12, 202, 255.
Messenger RNA, 234, 241, 248, 255, 264.

**N**

Nuclear Polyadenylation, 255.
Nucleases and Ligases, 147.
Nucleus, 107, 119.

**O**

Ovarian Cancer, 184, 185.
Oxygen Evolution, 68.

**P**

Penetration, 113.
Peroxisomes, 111.
Phospholipids, 106, 131.
Phosphorylation, 35, 288.
Polymerases, 148, 243.
Primer on RNA, 254.
Prokaryotic Cells, 104.
Prokaryotic cells, 5, 10, 105.
Protein Synthesis, 167.

**R**

Ribonucleic Acid, 119, 230.
Ribozyme, 250, 252, 253.
RNA Genomes, 236.
Rubisco, 77, 78, 80, 82, 83.
Rubisco Activase, 78.

**S**

Saccharides, 130.
Sea Urchins, 20.
Sex Chromosomes, 109.
Sexual Reproduction, 6.
Single Gene Disorder, 176.
Spectrophotometer, 63.
Splicing, 245, 249, 281.
Split Genes, 247.
Spores, 105.

**V**

Vacuoles, 111.
Variants, 2.
Viruses, 103, 111, 112, 113.
Visual Evidence, 245.
Vitamins, 122, 134.

**Z**

Zona Pellucida, 23.

❑❑❑